Science, Tech
and the Labo

Marxist Studies
volume I

edited by
Les Levidow and Bob Young

CSE BOOKS

Science, Technology and the Labour Process: Marxist Studies, Volume I was first published by CSE Books, 55 Mount Pleasant, London WC1X 0AE and in the USA by Humanities Press Inc., Atlantic Highlands, New Jersey 07716, in March 1981.

Cover Design by Tony Fry
Cover: Newton by William Blake (Tate Gallery)

British Library Cataloguing in Publication Data
Science, technology and the labour process.
Vol. 1
1. Labor and laboring classes
2. Technological innovations
3. Capitalism
I. Levidow, Les II. Young, Robert M.
331 HD4901
ISBN 0-906336-20-1
ISBN 0-906336-21-X Pbk
ISBN in the USA: 01-391-02297-0

Typeset by
Range Left Photosetters
30 Clerkenwell Close
London EC1

Printed by Blackrose Press, London.

CONTENTS

Acknowledgements

Thanks to Jane Pascoe and Cliff Harper for their help with production.

Credits for Reprints

Nathan Rosenberg on 'Marx as a Student of Technology' is reprinted by permission from *Monthly Review*, volume 28, number 3 (July/August 1976), pp.56-77.

Norman Diamond on 'The Politics of Scientific Conceptualization' is reprinted by permission from *Science for the People*, volume VIII, number 3 (May 1976), pp. 14-17, 40.

Credits for Illustrations

Cover: William Blake's *Newton*, reprinted by permission of The Tate Gallery, London.
p. vi: Cliff Harper
p.10: *The Encyclopedia Britannica*, 9th edition, Volume VI, 1877, p.498.
p.18: Andrew Ure, *The Philosophy of Manufactures* (1835), Cass, 1967.
p.22: *The Encyclopedia Britannica*, 8th edition, c.1853, Plate XXI.
p.36: Rene Taton (ed.), *The Beginnings of Modern Science*, Thames & Hudson, 1964, pp.62-63.
p.56: Cliff Harper, 78a Crofton Road, London SE5, tel. (01) 701-3090.
p.68: *Wechselwirkung* 1 (May 1979), p.34.
p.111: Ellen Armstrong, reprinted from *Science for the People*, Sept/Oct 1980.
pp.124, 130, 140, 162-3, 170: Poster Collective, Box 50, 182 Upper Street, London N1 tel 01 388 0182. Also available is a film on Grunwick, *Unfinished Business*.
p.141: reprinted by permission of *Which?* magazine, June 1977 issue, p.311.
p.171: *Newsline*.
p.173: Cartoon by Bill Sanderson, reprinted from *Time Out*, 4 April 1980, p.8.
p.176, 177, 179, 204: Poster Collective.
p.181: *Changing Role of S.E. Asian Women*, back cover. PO Box 4000D, Berkeley, CA 94704, USA.
p.193: *Wechselwirkung* 6 (August 1980), p.29.
p.194: Cartoon by TUGG, reprinted by permission from *PUNCH*, 5 November 1980.

Notes on Contributors

MIKE COOLEY is employed as a Senior Development Engineer in the aerospace industry and has played a leading role in the Lucas Aerospace Combine Shop Stewards Committee. He lectures widely in English and German at international conferences and universities in Europe, Australia and the USA.

NORMAN DIAMOND is Executive Director of the Pacific Northwest Labor College in Portland, Oregon, and is currently engaged in organizing around health, safety and the environment. He formerly taught political science at Antioch College, Ohio.

MIKE DUNCAN has worked in the computer industry since the early 1970s, when he left university following a humanities course. As a member of the CSE Microelectronics Group, he collaborated on the book they prepared for CSE Books.

LES LEVIDOW is a member of the Radical Science Journal collective and the BSSRS Politics of Energy Group. He also works around the Conference of Socialist Economists, and plays in 'The National Interest' and 'Diomedes Komnenos'.

NATHAN ROSENBERG, Professor of Public Policy at Stanford University, also works with the OECD and SSRC, and has published numerous books and articles on technology, industry and economic policy.

BOB YOUNG, a member of the Radical Science Journal collective, is also involved in the activities of the CSE. He has published many articles on science, technology and the labour process; he is currently working on a television series exploring the social relations of science and technology.

EDWARD YOXEN is a lecturer in the Department of Liberal Studies in Science, University of Manchester. He was involved with the BSSRS Genetic Engineering Group, and has written many articles on the history and social impact of molecular biology.

"We've decided to call it the neutron chip.
It eliminates jobs but leaves the factory intact."

Introduction

LES LEVIDOW and BOB YOUNG

For too long there has been an unholy alliance between orthodox Marxism, social-democracy and bourgeois culture on the role of science, technology and medicine in the mode of production. Although the emphases have varied, there has been a general consensus that science and technology are somehow above the class struggle. In saying that the emphases have varied, we mean that some people have argued that 'scientificity' is above the battle or even is itself a model for socialism. Others claim that science is neutral, but open to use and abuse. Still others continue to believe that science is pure but that technology, as the application of otherwise neutral science within the social order, is therefore shaped by society.

Of course, we could see nature as being 'out there', and hence see the scientific elucidation of it as independent of humanity. However, only as nature becomes socially mediated does it have any practical existence for humanity. All attempts to characterise nature, to know it and to shape it to human purposes are constituted by the prevailing social priorities of a given epoch. Historical forces are currently playing an important role in making nonsense of any approach which fails to acknowledge this.

It will come as no news to our readers to be told that capitalism is currently restructuring around science, technology and medicine. This has been proclaimed in so many newspaper and magazine articles, and in so many television programmes, that it is tempting for us to be suspicious and to believe that it therefore can't be true. The air of 'Gee whiz' which has accompanied, in particular, media presentations of the chip, could tempt the sceptical Marxist into believing that there is more hype than substance in the claims which are being made.

Of course, there is undoubtedly much hype in the claims made for an almost limitless trend towards replacing workers with machines, replacing living labour with dead labour. For such a trend inescapably aggravates capital's present difficulties in extracting surplus value from living labour. Furthermore, any technological change entails sober calculations as to whether it is marginally worthwhile to mechanize (or automate) a particular segment of a labour process.

Although the expected loss of jobs certainly does present a grave threat to the working class, speculation over the number of jobs to be lost has tended to obscure the qualitative trend as to the *kinds* of jobs being lost and as to the regimentation of the remaining or newly created jobs. Capital's priority for automation is to attack those stages of the productive cycle which have the most space for workers to hold down their own pace of work, those sites with the greatest 'porosity of labour' — be it clerical offices, paint, trim & assembly shops, or stock rooms. It is this selectivity which defines the threat posed to working class collective power by restructuring.

Furthermore, these transformations extend far beyond questions of productivity, far beyond the point of production, extending to whole areas of life not previously invaded by commodity exchange or state control. If one adds together the impact of growing specialisation, the handing over of domains to experts, the re-constituting of conception, gestation, disease, repair and even death . . . and if one connects these biomedical changes with the re-founding via biotechnology of agriculture, husbandry, food processing, mining, petroleum extraction, fuels, building materials, chemistry . . . and if one connects these with the all-too-familiar microelectronics, especially its impact on information, communication, surveillance, pacing and control . . . then one begins to get some sense of the depth and breadth of the restructuring which is going on. Hype or not, this is a list of capital's intentions. Whether or not they will be fulfilled is a question of the contingencies of investment, nature's refractoriness, the vicissitudes of research and development, and — of course — class struggle.

In this series of volumes we are attempting to bring together a number of studies which juxtapose Marxist analyses of science, technology and medicine with those which have been developed in other spheres. In so doing, it would be both disingenuous and premature for us to suggest that an integrated theory of the existing — much less the desirable — relationship between science, technology and medicine and the social order has already been elaborated. The aim of these volumes is, rather, to bring to our readers' attention both key texts and newly commissioned studies, just as previous CSE publications have already done in earlier volumes presenting key texts (on *The Labour Process & Class Strategies* and *Technology & Toil*) and new studies (on *Value* and *Microelectronics*). Our aims are to bring into question the privileged status of science, technology and medicine in the left as well as to help open up both the content and social relations of these domains for political struggle.

Although we think that the articles are clear and accessible, here we will add some of our reasons for having chosen this particular collection for our first volume.

Nathan Rosenberg's article on 'Marx as a Student of Technology' was very helpful to the CSE Labour Process/Left Strategy Group in the early stages of its attempts to make connections between Marxism and the critique of science. It refutes readings of Marx which present him as a technological determinist, and it emphasises the central role of the contradiction between the forces and relations of production in the capitalist mode of production. Our reading of this article helped us to keep constantly in mind that machines are particular instances of frozen labour.

In 'The Politics of Scientific Conceptualisation', Norman Diamond conveys the same insight about scientific knowledge. Choosing his examples from the scientific revolutions of the 16th and 17th centuries, he drives home the point that the most abstract theoretical work in science frames nature in terms which are characteristic of the world view of that period. This does not make the knowledge false or merely relative, but it does help explain why people approach nature in the way that they do in a given period. Furthermore his critical method can help reveal the social relations which present-day scientific concepts both presuppose and re-inforce.

In 'The Taylorisation of Intellectual Work', Mike Cooley takes us directly to concrete applications of science and technology for gaining greater control over the labour process. The sphere to which he draws attention is not that of routinised manual work but of supposedly creative intellectual work. This is the latest chapter of the history of which Marx wrote when he said,

> It would be possible to write a whole history of the inventions made since 1830 for the sole purpose of providing capital with weapons against working class revolt (*Capital*, Vol. I, Penguin, p. 563).

Cooley also describes struggles against these developments which have led workers to realise that unless they intervene in the origination of new technologies, and unless they take control over research, development and the labour process of production, they will be reduced to a holding action in the face of a scientific management advancing into white-collar work. This is a perspective which he develops further in *Architect or Bee?*, and which is being followed in *CAITS Quarterly*.

Edward Yoxen's article, 'Life as a Productive Force', complements Cooley's work in a way which is analogous to the relationship between the essays by Rosenberg and Diamond, i.e., the relationship between capitalist technology and the bourgeois scientific world-view. Yoxen looks at the fundamental theoretical changes in the conception of 'life' which have led from molecular biology to genetic engineering. A recent joint report by a committee of the Royal Society and various government bodies has argued that bio-technology will be as characteristic of the 21st century as chemistry and electricity have been of the 20th. This essay works carefully through the various levels of the politics of this scientific conceptualisation, which is proving at least as important in our epoch as the Copernican Revolution was in its epoch. It also shows the diminishing mediations between a conceptual re-definition of nature — in this case, as information, communication and control — on the one hand, and direct commercial exploitation, on the other.

In 'Grunwick: The Social Contract Meets the 20th Century Sweatshop', Les Levidow is as explicit about the problems of organised working class struggle as Yoxen is reticent. But the main focus of his article is the interface between high technology and exploitative working conditions. Similarly, while Yoxen shows how deep is the ideological constitution of concepts of nature in the life sciences, so the articulations between the Grunwick photo processing plant and its consumer market demonstrate the great extent of the ideological constitution of the way that snapshots represent people's experiences by providing instant nostalgia of colourful holidays.

Lastly, in 'Microelectronics: Five Areas of Subordination',

Mike Duncan provides various cases where the 'Gee whiz' technology of the microprocessor, far from freeing people from meaningless toil, sharply increases real subordination in pacing, surveillance and sweatshop conditions. This occurs in the manufacture as well as in the applications of the 'Mighty Micro'. This article complements the class analysis of the new technology which he and others have already presented in *Microelectronics: Capitalist Technology and the Working Class*. We hope that the inclusion of his article here will place that book in a much wider perspective on science and technology in general.

In future volumes we hope to provide further combinations of specially commissioned and reprinted essays which will illuminate different areas as well as different levels of the political economy, social relations and content of science, technology and medicine. In particular, we hope to provide more concrete historical analyses bearing closely on the current re-thinking of orthodox Marxism in the light of capital's latest strategies.

According to the scientistic Marxism inherited from the 2nd and 3rd Internationals, the interests of the working class are bound up with the project of 'developing the forces of production' — be it by sharing the benefits made available as capitalism did so, or by establishing socialism in order to surpass capitalism's limited ability to do so. However, while capitalist development has tended to increase the productivity of living labour, it has also re-organised the labour process to overcome worker insubordination. Furthermore, its latest developments (biotechnology, automation, nuclear power) put politics more firmly in command by aiming at a more directly political control over the production and reproduction of daily life, extending methods of factory discipline into the state's management of the social totality.

As the nature of the 'benefit' shifts more specifically to managerial tyranny for its own sake — almost indifferent to productive forces as such — the traditional question of how to share the benefits becomes at best meaningless, at worst misleading. The new question becomes how to subvert and supersede capital's project of degrading the working class's own productive forces.

It would be politically regressive, then, to analyse the labour process through a ready-made Marxism which looks to liberate the newest forms of fixed capital from capitalist social relations which contain and restrict them. Our task, rather, is to take up the capitalist labour process as a challenge for revitalising Marxism, making it adequate to the project of liberating living labour from the increasingly destructive direction of capitalist 'development'.

The perspective which underlies this volume is one which is beginning to be worked out as a result of the mutual influence of the development of a 'labour process perspective' within the CSE and tendencies within the radical science movement which have moved on from posing the science/ideology question to analysing science as a labour process. Readers who wish to look further into this developing perspective should consult the *Radical Science Journal*, the special issue of *Science for People* (no. 43/44) on 'Science Under Capitalism', and Mike Hales' book on *Living Thinkwork*. Attempts have already been made to express this synthesis in a more general form in two articles: 'The Problem of Articulation in Left Strategies' by the CSE Labour Process/Left Strategy Group, and 'Science, Technology, Medicine and the Socialist Movement' by the RSJ Collective.

We hope that this volume is an example of stronger links between the radical science movement and the Marxist analysis of apparently 'non-scientific' aspects of society. We also hope that it will help to increase the sophistication with which scientists approach politics and political activists approach science. Perhaps the day will come when we no longer find ourselves moving uneasily between arguments with people who still insist that 'Marxism = science' and our own secret fear that scientism is genetic.

Suggested Reading List

Stanley Aronowitz, 'Marx, Braverman and the Logic of Capital', *Insurgent Sociologist* VIII, 2/3 (Fall 1978): Essays on the Social Relations of Work & Labor, 126-46.

CAITS Quarterly, c/o NELP, Longbridge Road, Dagenham, Essex.

Simon Clarke, 'Marxism, Sociology and Poulantzas' Theory of the State', *Capital & Class* 2 (Summer 1977), 1-31.

CSE Labour Process/Left Strategy Group, 'The Problem of Articulation in Left Strategy', *CSE Conference Papers*, 1979, 101-18.

Mike Cooley, *Architect or Bee?*, Langley Technical Services, 1980.

David Dickson, 'Science and Political Hegemony in the 17th Century', *RSJ* 8 (1979), 7-38.

Mike Hales, 'Down With the Consumption of Knowledge — An Open Letter to Academics' and 'Thinkers and Doers — Intellectuals and the Politics of Work: A Review of Philip Kraft', *RSJ* 9 (1979), 5-13, 111-25.

Mike Hales, *Living Thinkwork*, CSE Books, 1980.

Mark Harrison, review of *Socialist Construction and Marxist Theory: Bolshevism and Its Critique*, in *Capital & Class* 8 (Summer 1979), 148-52.

The Labour Process & Class Strategies (CSE Pamphlet No. 1), Stage 1, 1976.

Antonio Negri, 'Domination and Sabotage', in *Working Class Autonomy and the Crisis*, CSE Books, 1979, pp. 93-117 (especially Chapter 7, 'Regarding Forces of Production', 119-23).

Raniero Panzieri, 'Surplus Value and Planning', in *The Labour Process and Class Strategies*, Stage 1, 1976, pp. 4-25.

Raniero Panzieri. 'The Capitalist Use of Machinery: Marx Versus the "Objectivists" ', in *Outlines of a Critique of Technology*, Inklinks, 1980, pp. 44-68.

RSJ 6/7 (1978) – Special Labour Process Issue, includes 'Reading Lists in Marxism & Science, the Labour Process and Psychoanalysis', 144-50.

RSJ Collective, 'Science, Technology, Medicine and the Socialist Movement', forthcoming.

Bob Young, review of Braverman, *RSJ* 4 (1976), 81-93.

Bob Young, 'Science *is* Social Relations', *RSJ* 5 (1977), 65-131.

Bob Young, 'Science Is A Labour Process', *Science For People* 43/44: special issue on Science Under Capitalism, 31-37.

Marx as a Student of Technology

NATHAN ROSENBERG

I

This paper will attempt to demonstrate that a major reason for the fruitfulness of Marx's framework for the analysis of social change was that Marx was, himself, a careful student of technology. By this I mean not only that he was fully aware of, and insisted upon, the historical importance and the social consequences of technology. That much is obvious. Marx additionally devoted much time and effort to explicating the distinctive characteristics of technologies, and to attempting to unravel and examine the inner logic of individual technologies. He insisted that technologies constitute an interesting subject, not only to technologists, but to students of society and social pathology as well, and he was very explicit in the introduction of technological variables into his arguments.

I will argue that, quite independently of whether Marx was right or wrong in his characterization of the future course of technological change and its social and economic ramifications, his formulation of the problem still deserves to be a starting point for any serious investigation of technology and its ramifications. Indeed, the following statement by Marx, amazingly fresh over a century later, reads like a prolegomenon to a history of technology which still remains to be written:

> A critical history of technology would show how little any of the inventions of the eighteenth century are the work of a single individual. Hitherto there is no such book. Darwin has interested us in the history of Nature's Technology, i.e., in the formation of the organs of plants and animals, which organs serve as instruments of production for sustaining life. Does not the history of the productive organs of man, of organs that are the material

> basis of all social organization, deserve equal attention? And would not such a history be easier to compile since, as Vico says, human history differs from natural history in this, that we have made the former, but not the latter? Technology discloses man's mode of dealing with Nature, the process of production by which he sustains his life, and thereby also lays bare the mode of formation of his social relations, and of the mental conceptions that flow from them.[1]

In what follows, I will focus first upon Marx's alleged technological determinism, second on his views on the characteristics of Modern Industry which are responsible for its high degree of technological dynamism, and finally on the special importance which Marx attaches to the role of the capital-goods sector in the generation of technological change. An important question regarding Marx's treatment of technology is, quite simply, what it was about his method or approach which made him so much more perceptive on this subject than any of his contemporaries. I think a few tentative methodological observations, upon which I will elaborate later, may be useful at the outset.

The method of historical materialism which Marx utilized was one which emphasized the interactions and conflicts of social classes and institutions, not individuals. Thus for Marx invention and innovation, no less than other socioeconomic activities, were best analyzed as social processes rather than as inspired flashes of individual genius. The focus of Marx's discussion of technological change is thus not upon individuals, however heroic, but upon a collective, social process in which the institutional and economic environments play major roles.

Marx's historical approach was one which emphasized the discontinuous nature of social evolution, an evolutionary process which was, for him, moving forward under capitalism, just as it had under earlier forms of social organization. Rather than viewing capitalism as the final, logical outcome of a smooth, lengthy evolutionary process, Marx treated it as simply one stage in the process of historical evolution, while looking for the unique features of the capitalist forces of production and attempting to understand them in dynamic terms.

I would argue that the dialectical method is the most important factor in understanding the methodological basis for Marx's unique insights. Rather than positing some unidirectional chain of causation for technological change, Marx offers a far

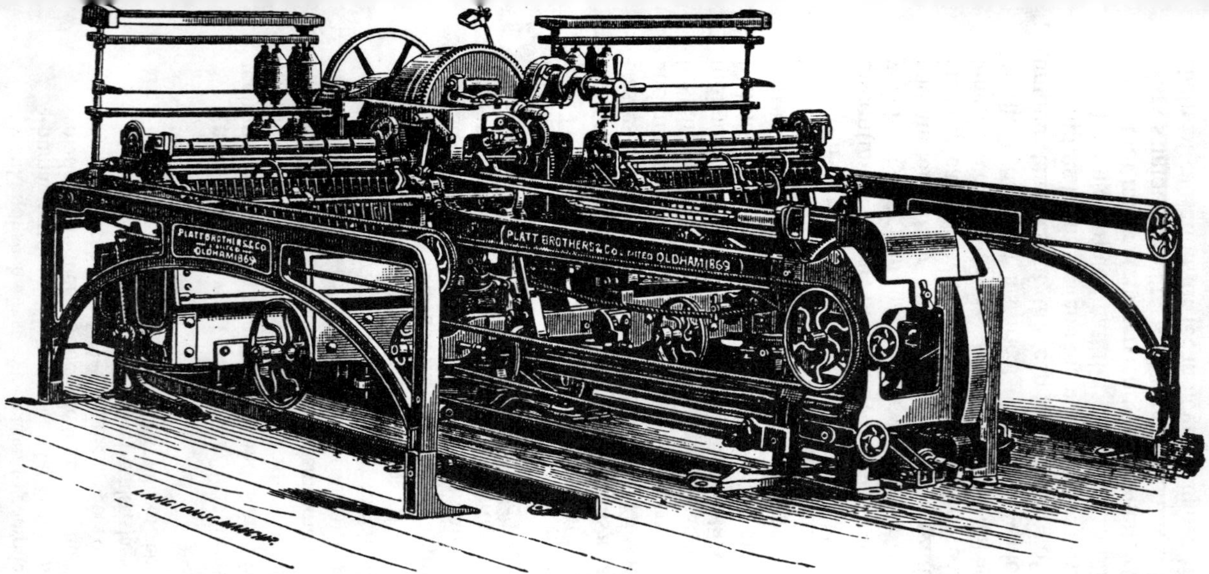

Self-acting Mule

It would be possible to write a whole history of the weapons made since 1830 for the sole purpose of providing capital with weapons against working class revolt. We would mention, above all, the self-acting mule, because it opened up a new epoch in the automatic system . . . Of the invention of the self-acting mule, Ure says: 'A creation designed to restore order among the industrious classes . . .'

Karl Marx, *Capital*, Volume I.

richer mode of analysis, one which emphasizes the mutual interactions and feedbacks between economy and technology. His analysis of the rise of the system of 'machinofacture' and its implications for technological change, which I will examine in this paper, is an important example of the insights yielded by such a method.

II

First of all, it is necessary to advert briefly to the oft-repeated view that Marx was a technological determinist.[2] If by this we mean that technological forces are the decisive factor in generating socioeconomic changes — that technological factors are, so to speak, the independent variable in generating social change, which constitutes the dependent variable — it is easy to demonstrate that Marx subscribed to no such simplistic view. No doubt certain passages can be cited to support such an interpretation — most notably, of course, the statement from *The Poverty of Philosophy* that 'The handmill gives you society with the feudal lord; the steam-mill, society with the industrial capitalist.'[3] But such an interpretation, relying upon a few such aphoristic assertions, often tossed out in the heat of debate (as the quotation from *The Poverty of Philosophy* was thrown out in criticizing Proudhon) finds little support in Marx's own treatment of the major historical episodes with which he was concerned. Indeed, in a process no less central to Marx than the historical rise of capitalism itself, technological factors play no immediate role at all. For Marx, capitalism developed in Western Europe basically in response to growing markets and related opportunities for profit-making associated with the geographic explorations of the fifteenth century. The *locus classicus* for this view is, of course, the opening pages of the *Communist Manifesto*:

> The discovery of America, the rounding of the Cape, opened up fresh ground for the rising bourgeoisie. The East-Indian and Chinese markets, the colonization of America, trade with the colonies, the increase in the means of exchange and in commodities generally, gave to commerce, to navigation, to industry, an impulse never before known, and thereby, to the revolutionary element in the tottering feudal society, a rapid development.
>
> The feudal system of industry, under which industrial production was monopolized by closed guilds, now no longer

> sufficed for the growing wants of the new markets. The manufacturing system took its place. The guild-masters were pushed on one side by the manufacturing middle class; division of labor between the different corporate guilds vanished in the face of division of labor in each single workshop.
>
> Meantime the markets kept ever growing, the demand ever rising. Even manufacture no longer sufficed. Thereupon, steam and machinery revolutionized industrial production. The place of manufacture was taken by the giant, Modern Industry, the place of the industrial middle class, by industrial millionaires, the leaders of whole industrial armies, the modern bourgeois.
>
> Modern industry has established the world-market, for which the discovery of America paved the way. This market has given an immense development to commerce, to navigation, to communication by land. This development has, in its turn, reacted on the extension of industry; and in proportion as industry, commerce, navigation, railways extended, in the same proportion the bourgeoisie developed, increased its capital, and pushed into the background every class handed down from the Middle Ages.[4]

I have taken the liberty of quoting at length from a familiar source because this passage is, it seems to me, a definitive refutation of the view that Marx was a technological determinist. The question is: What are the factors which *initiate* change? What are the factors which cause other factors to change? A resolute technological determinist would presumably argue that the entire process of European expansion was initiated by navigational improvements which in turn generated the growing of overseas markets. But this is clearly not Marx's own view here. In fact, Marx states *twice* that the improvements in navigation were caused by the prior growth in markets and commercial opportunities. The passage makes it unmistakably clear that the technological changes associated with the two stages of capitalist development — the manufacturing system and Modern Industry — were responses to an expanding universe of profit-making opportunities.

For Marx, then, capitalist relationships emerged when the growth of profit-making opportunities led to an expansion in the size of the productive unit beyond that which was characteristic of the medieval craft workshop. The mere quantitative expansion of such workshops led eventually to qualitative changes of a most basic sort in social relationships.[5] Although the system of manufacture totally dominated the first two and a half centuries of Western capitalism and led to major transformations in social relationships,[6] it was not associated with any major technological innovations. 'With regard to the mode of production itself, manufacture, in its strict meaning, is hardly to be distinguished, in its earliest stages, from the handicraft trades of the guilds, otherwise than by the greater number of workmen simultaneously employed by one and the same individual capital. The workshop of the mediaeval master handicraftsman is simply enlarged.'[7]

To regard Marx as a technological determinist, then, is tantamount to ignoring his dialectical analysis of the nature of historical change.[8] The essence of this view is that the class struggle, the basic moving force of history, is itself the product of fundamental contradictions between the forces of production and the relations of production. At any point in historical time, new productive forces emerge, not exogenously or as some mysterious *deus ex machina*, but rather as a dialectical outcome of a larger historical process in which *both* the earlier forces and relations of production play essential roles. As Marx forcefully put it: 'It must be kept in mind that the new forces of production and relations of production do not develop out of *nothing*, nor drop from the sky, nor from the womb of the self-positing Idea; but from within and in antithesis to the existing development of production and the inherited, traditional relations of property.'[9]

Thus, for Marx the basic rhythm of human history is the outcome of this dialectical interaction between the forces and the relations of production. To categorize Marx as a technological determinist one would have to demonstrate first that he does *not* intend his historical argument to proceed in a dialectical form. I think it is easy to demonstrate that he does.

III

If Marx was not a technological determinist he did, nonetheless, attach great importance to technological factors. The reasons for this are made clear in Chapter 7 of the first volume of *Capital*. Technology is what mediates between man and his relationship with the external, material world. But in acting upon that material world, man not only transforms it for his own useful purposes (that is to say, 'Nature becomes one of the organs of his activity'[10]) but he also, unavoidably, engages in an act of self-transformation and self-realization. 'By thus acting on the external world and changing it, he at the same time changes his own nature.'[11] Technology, therefore, is at the center of those activities which are distinctively human. For technology comprises those instruments which determine the effectiveness of man's pursuit of goals which are shaped not only by his basic instinctive needs, but also those formulated and shaped in his own brain. 'A spider conducts operations that resemble those of a weaver, and a bee puts to shame many an architect in the construction of her cells. But what distinguishes the worst architect from the best of bees is this, that the architect raises his structure in imagination before he erects it in reality. At the end of every labor process, we get a result that already existed in the imagination of the laborer at the commencement.'[12]

The informed student of society, therefore, can infer much concerning the nature of a society, its intellectual attainments, its organization, and its dominant social relationships, by studying the instruments of human labor. But again it needs to be insisted that Marx's position here cannot be reduced to a crude technological determinism. In a highly perceptive passage which is sometimes cited as evidence of his technological determinism, Marx is, in fact, pointing to what can be *inferred* about the nature of earlier societies from their remaining artifacts. 'Relics of by-gone instruments of labor possess the same importance for the investigation of extinct economical forms of society, as do fossil bones for the determination of extinct species of animals. It is not the articles made, but how they are made, and by what instruments, that enables us to distinguish different economical epochs. Instruments of labor not only supply a *standard* of the degree of development to which human labor has attained, but they are also *indicators* of the social conditions under which that labor is carried on.'[13] I believe this passage is evidence of technological determinism in about the same sense that one is entitled to say that a thermometer *determines* body temperature or a barometer *determines* atmospheric pressure. All such statements are equally specious and misleading in mixing up measurement with causation.

The decisive technological changes with which Marx is concerned begin around the middle of the eighteenth century. It is at this point that Britain began her transition from an industrial system of manufacture to what Marx calls Modern Industry. The general outline of his analysis, as presented in Chapters 13–15 of *Capital*, is well known and need not be repeated here. Rather, I will focus on certain specific features which, I believe, are still insufficiently appreciated.

First of all, Marx posed and dealt with a basic question concerning the nature of technology which has never received the attention it deserves. It is widely accepted that modern capitalist societies have achieved high levels of productivity because of the systematic application of scientific knowledge to the productive sphere. As Kuznets has stated: 'The epochal innovation that distinguishes the modern economic epoch is the extended application of science to problems of economic production.'[14] Our awareness of the importance of modern science (an awareness which is not, of course, confined to its purely economic significance) has led to a mushrooming interest in the history of science, which is now a thoroughly respectable academic discipline. But although we study the history of science (in some cases with the financial support of the National Science Foundation), the study of the history of technology is still largely (although by no means entirely) neglected. And yet to the extent that we are interested in the economic importance of science, we need to study the history of technology, because not all technologies will permit, or will permit in equal degrees, the *application* of scientific knowledge to the productive sphere. The growth of science, *by itself*, is not a sufficient condition for the growth of productivity. To believe that it is, is to ignore the mediating role of technology between man and nature. It was one of Marx's most important accomplishments to have posed precisely this question: What are the characteristics of technologies which make it possible to apply scientific knowledge to the productive sphere? Moreover, I think Marx suggested an answer which was quite adequate for his own historical period, given the nature of the industrial technology of his day and the state of scientific development. Still, this immensely important question needs to be posed and studied anew for the far more complex technologies, as well as the far more sophisticated bodies of scientific knowledge, which have emerged in the century since Marx wrote *Capital*. From the vantage point of the mid-nineteenth century, Marx's answer ran along the following lines.

The manufacturing system, which was the dominant mode of production of early capitalism, developed a high degree of worker specialization. Whereas the medieval handicraftsmen performed a whole range of operations in the production of a single commodity, the manufacturing system broke down the productive process in a series of discrete steps, and assigned each step to a separate detail laborer. However, although this growing specialization of work had highly significant consequences with which Marx was intensely concerned, it nevertheless shared a basic feature in common with the medieval handicraft system: a continued reliance upon human skills and capacities.

> Whether complex or simple, each operation has to be done by hand, retains the character of a handicraft, and is therefore dependent upon the strength, skill, quickness, and sureness, of the individual workman in handling his tools. The handicraft continues to be the basis. This narrow technical basis excludes a really scientific analysis of any definite process of industrial production, since it is still a condition that each detail process gone through by the product must be capable of being done by hand and of forming, in its way, a separate handicraft. It is just because handicraft skill continues, in this way, to be the foundation of the process of production, that each workman becomes exclusively assigned to a partial function, and that for the rest of his life, his labor power is turned into the organ of this detail function.[15]

Although, therefore, the manufacturing system achieved a growth in productivity through the exploitation of a new and more extensive division of labor, a rigid ceiling to the growth in productivity continued to be imposed by limitations of human strength, speed, and accuracy. Marx's point, indeed, is more general: science itself can never be extensively applied to the productive process so long as that process continues to be dependent upon forces the behavior of which cannot be predicted and controlled with the strictest accuracy. Science, in other words, must incorporate its principles in impersonal machinery. Such machinery may be relied upon to behave in accordance with scientifically established physical relationships. Science, however, cannot be incorporated into technologies dominated by large-scale human interventions, for human action involves too much that is subjective and capricious. More generally, human beings have wills of their own and are therefore too refractory to constitute reliable, i.e., controllable, inputs in complex and interdependent productive processes.

The decisive step, then, was the development of a machine technology which was not heavily dependent upon human skills or volitions, where the productive process was broken down into a series of separately analyzable steps. The historic importance of the manufacturing system was that it had provided just such a breakdown. The historic importance of Modern Industry was that it incorporated these separate steps into machine processes to which scientific knowledge and principles could now be routinely applied. 'The principle, carried out in the factory system, of analyzing the process of production to its constitutent phases, of solving the problems thus proposed by the application of mechanics, of chemistry, and of the whole range of the natural sciences, becomes the determining principle everywhere.'[16] When this stage has been reached, Marx argues, technology becomes, for the first time, capable of indefinite improvement:

> Modern industry rent the veil that concealed from men their own social process of production, and that turned the various, spontaneously divided branches of production into so many riddles, not only to outsiders, but even to the initiated. The principle which it pursued, of resolving each process into its constituent movements, without any regard to their possible execution by the hand of man, created the new modern science of technology. The varied, apparently unconnected, and petrified forms of the industrial process now resolved themselves into so many conscious and systematic applications of natural science to the attainment of given useful effects. Technology also discovered the few main fundamental forms of motion, which despite the diversity of the instruments used, are necessarily taken by every productive action of human body; just as the science of mechanics sees in the most complicated machinery nothing but the continual repetition of the simple mechanical powers.[17]

I suggest that Marx's insight into the historical interrelationships between science and technology was extraordinarily perceptive and that it ought to be treated as a starting point for the vastly more complex interrelationships which have characterized the last century of capitalist development.[18]

IV

In the remainder of this paper I propose to concentrate upon Marx's analysis of the unique role and importance of the capital-goods sector. Although Marx is generally recognized as the father of the two-sector model, this recognition has been confined primarily to the usefulness of such models in the explanation of the inherent instability of capitalist economies, after the fashion in which Marx employed such models in the second volume of *Capital*. His work, however, suggests much more than this. The identification and isolation of a capital-goods producing sector offers rich possibilities for the further

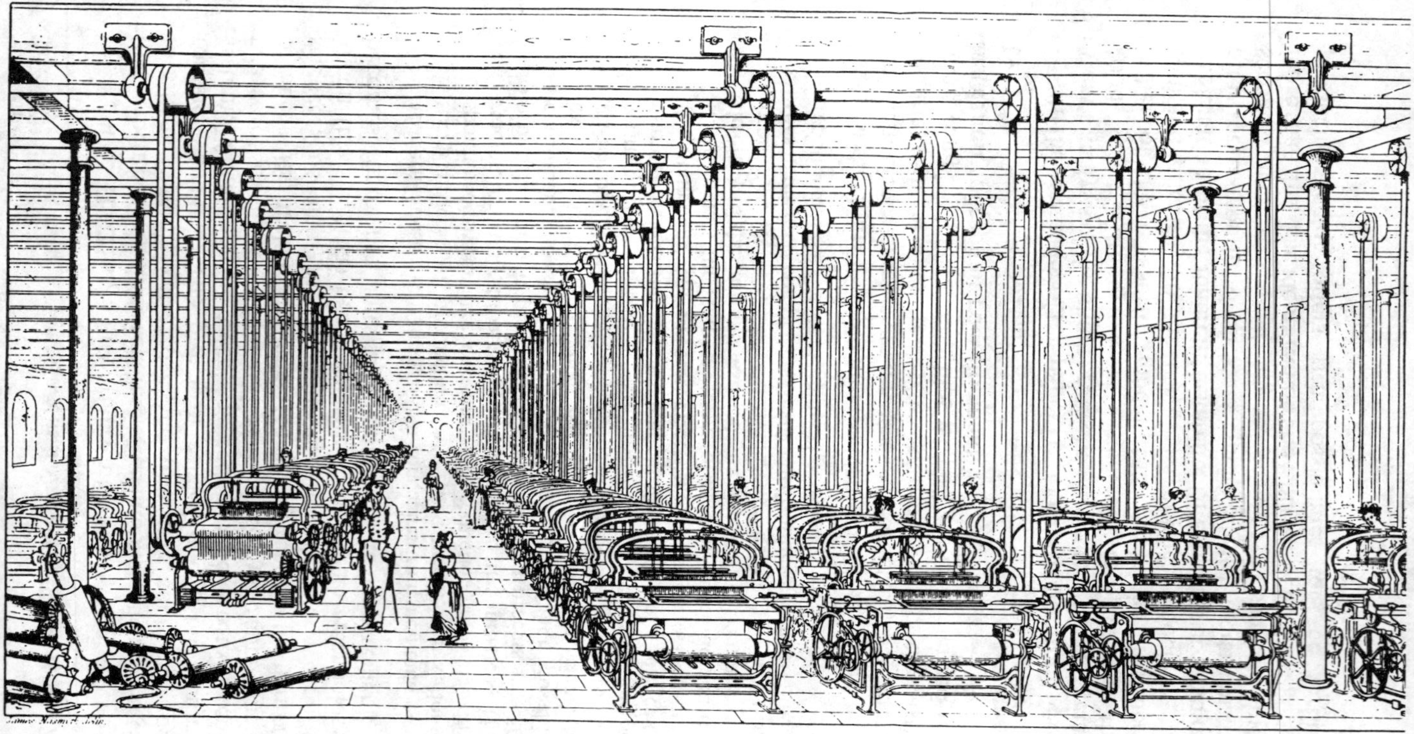

POWER LOOM FACTORY OF THOMAS ROBINSON ESQ.[r]
STOCKPORT.

understanding of the form and the mechanism of diffusion of technological change, as well as other critical aspects of the behaviour of capitalist societies.

In the early stages of the development of Modern Industry, machinery was produced by handicraft and manufacturing methods.[19] Although such methods were sufficient in the early stages, the growth in machine size and complexity and requirements for improvement in machine design and performance characteristics created demands which proved to be incompatible with the limited capacities of handicraft and manufacturing technologies. Indeed, 'Such machines as the modern hydraulic press, modern powerloom, and the modern carding engine, could never have been furnished by Manufacture.'[20] The realization of the full productive possibilities of Modern Industry therefore required that machine techniques be employed in the construction of the machines themselves. This is the final stage in the 'bootstrap' operation, the stage by which Modern Industry completes its liberation from the constraints of the old technology. 'Modern Industry had therefore itself to take in hand the machine, its characteristic instrument of production, and to construct machines by machines. It was not till it did this, that it built up for itself a fitting technical foundation, and stood on its own feet. Machinery, simultaneously with the increasing use of it, in the first decades of this century, appropriated, by degrees, the fabrication of machines proper.'[21]

Once this stage of technological maturity has been attained, modern capitalism may be regarded as being in full possession of those extraordinary technological means which sharply distinguish it from all earlier stages in the development of man's productive capacities:

> So soon . . . as the factory system has gained a certain breadth of footing and a definite degree of maturity, and, especially, so soon as its technical basis, machinery, is itself produced by machinery; so soon as coal mining and iron mining, the metal industries, and the means of transport have been revolutionized; so soon, in short, as the general conditions requisite for production by the modern industrial system have been established, this mode of production acquires an elasticity, a capacity for sudden extension by leaps and bounds that finds no hindrance except in the supply of raw materials and the disposition of the produce.[22]

It was one of Marx's enduring accomplishments that he was

among the first to perceive the inevitability of the trend towards bigness. This perception was, again, firmly rooted in his careful study of technological forces at work in mid-nineteenth century British capitalism. Marx asserted in Chapter 25 of the first volume of *Capital* ('The General Law of Capitalist Accumulation') the decisive economic advantages of capitalist production on a large scale. The nature of these advantages is carefully categorized and analyzed, and numerous specific examples are presented in Chapter 5 of the third volume ('Economy in the Employment of Constant Capital').

When, as a result of the process of capital accumulation, the capitalist economy has acquired a sufficiently large complement of capital goods, and therefore also a well-defined sector devoted to the production of capital goods, the system at this stage acquires a new source of productive dynamism. First of all there are the opportunities, when the scale of production is sufficiently large, for the exploitation of what we have come to call indivisibilities:

> In a large factory with one or two central motors the cost of these motors does not increase in the same ratio as their horse power and, hence, their possible sphere of activity. The cost of the transmission equipment does not grow in the same ratio as the total number of working machines which it sets in motion. The frame of a machine does not become dearer in the same ratio as the mounting number of tools which it employs as its organs, etc. Furthermore, the concentration of means of production yield a saving on buildings of various kinds not only for the actual workshops, but also for storage, etc. The same applies to expenditures for fuel, lighting, etc. Other conditions of production remain the same, whether used by many or by few.[23]

Furthermore, when production takes place on a sufficiently large scale, it eventually becomes worthwhile to take steps to utilize waste, or by-product, materials. 'The general requirements for the reemployment of these 'excretions' are: large quantities of such waste, such as are available only in large-scale production, improved machinery whereby materials, formerly useless in their prevailing form, are put into a state fit for new production; scientific progress, particularly chemistry, which reveals the useful properties of such waste.'[24]

Most generally, the existence of a large stock of capital goods now provides powerful economic incentives for innovations of a capital-saving nature. That is to say, there now exist great opportunities for increasing the rate of profit 'by reducing the value of the constant capital required for commodity production.'[25] This involves not only the development of improved machinery — steam engines which deliver a greater amount of power with the same expenditure of capital and fuel — but also *technological change in the machinery-producing sector itself.* As a result of such improvements in the machine-building sector, 'although the value of the fixed portion of constant capital increases continually with the development of labor on a large scale, it does not increase at the same rate.'[26] At this most advanced stage of capitalist development, inter-industry relationships come to play a most important role, because the rate of profit in one industry now becomes a function of labor productivity in another industry. 'What the capitalist thus utilizes are the advantages of the entire system of the social division of labor. It is the development of the productive power of labor in its exterior department, in that department which supplies it with means of production, whereby the value of the constant capital employed by the capitalist is relatively lowered and consequently the rate of profit is raised.'[27]

From an even broader perspective, the rate of profit may be increased by any capital-saving innovations, in whatever form. A large class of such innovations will therefore include all measures which reduce the turnover period of capital. From an economy-wide point of view, this has been precisely the effect of the communications revolution, one of the effects of which was a drastic reduction in the requirements for circulating capital. Marx deals with this source of capital-saving in Chapter 4 of the third volume of *Capital* ('The Effect of the Turnover on the Rate of Profit'):

> The chief means of reducing the time of circulation is improved communications. The last 50 years have brought about a revolution in this field, comparable only with the industrial revolution of the latter half of the eighteenth century. On land the macadamized road has been displaced by the railway, on sea the slow and irregular sailing vessel has been pushed into the background by the rapid and dependable steamboat line, and the entire globe is being girdled by telegraph wires. The Suez Canal has fully opened East Asia and Australia to steamer traffic. The time of circulation of a shipment of commodities to East Asia, at least twelve months in 1847 . . ., has now been reduced to almost as

Manufacture is a word, which, in the vicissitude of language, has come to signify the reverse of its intrinsic meaning, for it now denotes every extensive product of art, which is made by machinery, with little or no aid of the human hand; so that the most perfect manufacture is that which dispenses entirely with manual labour.

Andrew Ure, *The Philosophy of Manufactures* (1835)

The man of knowledge and the productive labourer come to be widely divided from each other, and knowledge, instead of remaining the handmaid of labour in the hand of the labourer to increase his productive powers... has almost everywhere arrayed itself against labour systematically deluding and leading them [the labourers] astray in order to render their muscular power entirely mechanical and obedient.

W. Thompson, *An Inquiry into the Principles of the Distribution of Wealth* (1824)

> many weeks. The two large centers of the crises of 1825–1857, America and India, have been brought from 70 to 90 percent nearer to the European industrial countries by this revolution in transport . . . The period of turnover of the total world commerce has been reduced to the same extent, and the efficacy of the capital involved in it has been more than doubled or trebled. It goes without saying that this has not been without effect on the rate of profit.[28]

All this from someone who is often described as treating the process of technological innovation as if it were purely a labor-saving phenomenon!

One further point which deserves to be made here is that Marx does not regard the reliance upon a new power source as the crucial element in the development of machines. Indeed, as he points out, in many of the early machines man was himself the prime mover. The distinctive element, as our earlier discussion has suggested, is the transfer of the control over the tool out of human hands.[29] This transfer of control involves a quantum leap forward, since 'the number of tools that a machine can bring into play simultaneously is from the very first emancipated from the organic limits that hedge in the tools of a handicraftsman.'[30] Furthermore, modern technology has even, *mirabile dictu*, invented a substitute for the human hand itself in the form of Henry Maudsley's slide rest.[31] This simple but ingenious device, as Marx perceptively notes, replaces not any particular tool, 'but the hand itself.' In this sense the slide rest is a technological breakthrough fully comparable in importance to the steam engine.

In his discussion of technological change within the capital goods sector, Marx has many acute observations which are tossed out without further development. New inventions often contain inefficient design features at the outset because the inventor has not shaken himself totally free of an earlier technology which is being displaced and whose operating principles have been rendered irrelevant. Marx observes:

> To what an extent the old forms of the instruments of production influenced their new forms at first starting, is shown by, amongst other things, the most superficial comparison of the present powerloom with the old one, of the modern blowing apparatus of a blast-furnace with the first inefficient mechanical reproduction of the ordinary bellows, and perhaps more strikingly than in any other way, by the attempts before the invention of the present

> locomotive, to construct a locomotive that actually had two feet, which after the fashion of a horse, it raised alternatively from the ground. It is only after considerable development of the science of mechanics, and accumulated practical experience, that the form of a machine becomes settled entirely in accordance with mechanical principles, and emancipated from the traditional form of the tool that gave rise to it.[32]

It is a shame that posterity has been deprived of some sketch of the ill-fated, two-footed locomotive! But, more seriously, the last sentence can be read as a striking anticipation of some of the central ideas of Abbott Payson Usher, probably the most careful twentieth-century student of the history of technology.[33] For Marx is insistent that technology has to be understood as a social process. The history of invention is, most emphatically, not the history of inventors. Here, as in so many other realms, Marx's position cannot be understood without dealing with the basic methodological question: What is the most appropriate unit of analysis? His answer is that for questions pertaining to long-term changes in technology, the individual is *not* the appropriate unit. In this particular instance Marx is insisting that technological change cannot be adequately understood by examining the contributions of single individuals, even though he often acknowledges the noteworthy contributions of such individuals. Rather, one needs to examine the way in which larger social forces continually alter the focus of technological problems which require solutions. Within this framework one may then examine how the productive process has, in the past, shaped the development of scientific and technological knowledge and skills.[34] One is then in a position to explore the social process of problem formulation and eventual solution. In all of this, however, although individual human beings are, inevitably, the actors, the *dramatis personae* of the historical process, the actual unfolding of the plot turns upon the larger social forces which shape their actions. Within this larger framework it is possible to see the efforts and contributions of *numerous* individuals even though those twin bastions of individualism, the patent office and writers of history textbooks, require that the names of *single* individuals be written alongside the names of particular inventions. But what is really involved is a process of a cumulative accretion of useful knowledge, to which many people make essential contributions, even though the prizes and recognition are usually accorded to the one actor who happens to have been on the stage at a critical moment.

Usher would agree with Marx's stricture, quoted earlier, that 'a critical history of technology would show how little any of the inventions of the eighteenth century are the work of a single individual.' Usher's own analysis turns upon the study of problem formulation in dealing with technology, focusing especially upon the process which he describes as 'the setting of the stage.' It is the correct setting of the stage which makes possible the eventual act of insight leading to the solving of the problem (whether Watt's separate condenser or Bessemer's converter). As Usher points out: 'Our analysis . . . redefines the question. It is not necessary to explain the final act of insight; the task now consists in explaining how the stage is set to suggest the solution of the perceived problem.'[35] Moreover, this is not the end of the process, because the initial act of insight is likely to lead to crude and primitive solutions. 'The setting of the stage leads directly to the act of insight by which the essential solution of the problem is found. But this does not bring the process to an end. Newly perceived relations must be thoroughly mastered, and effectively worked into the entire context of which they are a part. The solution must, therefore, be studied critically, understood in its fullness, and learned as a technique of thought or action. This final stage can be described as critical revision.'[36]

I would regard the completion of Usher's stage of critical revision as bringing us to essentially the same point in the development of a technology that Marx had in mind in referring to the process by which 'the form of a machine becomes settled entirely in accordance with mechanical principles, and emancipated from the traditional form of the tool that gave rise to it. Marx was aware of the regular need for something resembling Usher's final stage of critical revision and attached considerable importance to it. Indeed there is implicit in Marx's analysis a kind of 'life cycle' in the development of new techniques of production. New machines, when first introduced, are usually economically inefficient for two quite distinct reasons. First of all, the initial model has not yet had the opportunity to be subjected to a rigorous examination of its operations from which methods for performance improvements can be expected to flow. Marx pointed to continual improvements

> which lower the use-value, and therefore the value, of existing

> machines, factory buildings, etc. This process has a particularly dire effect during the first period of newly introduced machinery, before it attains a certain stage of maturity, when it continually becomes antiquated before it has time to reproduce its own value. This is one of the reasons for the flagrant prolongation of the working time usual in such periods, for alternating day and night shifts, so that the value of the machine may be reproduced in a shorter time without having to place the figures for wear and tear too high. If, on the other hand, the short period in which the machinery is effective (its short life vis-à-vis the anticipated improvements) is not compensated in this manner, it gives up so much of its value to the product through moral depreciation that it cannot even compete with hand-labor.[37]

The early model, therefore, is recognized to have a short life expectancy — a high rate of 'moral depreciation' — and this expectation that it will shortly be swept away by the competition of improved models is an ever present consideration in the mind of the capitalist.

Eventually the new machine, having been subjected to a series of design improvements, assumes a relatively stabilized form, and at this point it becomes possible for the capital-goods sector to develop techniques for producing the machine more cheaply. This is where the capital-goods sector plays its critical role in the ongoing competitive process:

> After machinery, equipment of buildings, and fixed capital in general, attain a certain maturity, so that they remain unaltered for some length of time at least in their basic construction, there arises a similar depreciation due to improvements in the methods of reproducing this fixed capital. The value of the machinery, etc., falls in this case not so much because the machinery is rapidly crowded out and depreciated to a certain degree by new and more productive machinery, etc., but because it can be reproduced more cheaply. This is one of the reasons why large enterprises frequently do not flourish until they pass into other hands, i.e., after their first proprietors have been bankrupted, and their successors, who buy them cheaply, therefore begin from the outset with a smaller outlay of capital.[38]

Marx emphasized in several other places the high cost of the early machine models by comparison with the later ones.[39] His argument suggests a great deal, not only about the process through which a capitalist economy generates new techniques, but also about the speed with which new techniques will be spread throughout the economy.[40]

I do not propose to discuss the question whether Usher's work was influenced by Marx. That is not, in my view, terribly important. But I do want to insist upon the fruitfulness of certain ways of conceptualizing the technological process under capitalism which Marx suggested, but never developed beyond some precocious and suggestive hints. I want also to render my judgement that American students of Marx, in creating a mode of analysis which they call 'the Marxist tradition,' have not been faithful, in at least one important respect, to the mode of analysis initiated by Marx himself. For Marx, as I have argued, was a close student both of the history of technology and its newly emerging forms.

These strictures are only somewhat less applicable to the British Marxian tradition. For although it is true that the two leading British figures in the history of technology, J.D. Bernal and Joseph Needham, have frequently adverted to the strong Marxian component of their thinking, they have not, any more than others, followed Marx's hints or elaborated upon his insights concerning the development of modern industrial technology. This is a task which still remains to be undertaken.

Ure says this of the colouring machines used in calico printing: "At length capitalists sought deliverance from this intolerable bondage" (namely the terms of their contracts with the workers, which they saw as burdensome) "in the resources of science, and were speedily re-instated in their legitimate rule, that of the head over the inferior members."

Then, speaking of an invention for dressing warps, whose immediate occasion was a strike, he says: "The combined malcontents, who fancied themselves impregnably intrenched behind the old lines of division of labour, found their flanks turned and their defences rendered useless by the new mechanical tactics, and were obliged to surrender at discretion." Of the invention of the self-acting mule, he says: "A creation destined to restore order among the industrious classes . . . This invention confirms the great doctrine already propounded, that when capital enlists science into her service, the refractory hand of labour will always be taught docility."

— Karl Marx, *Capital*, Volume I

Acknowledgements:

In preparing this paper, I have had the considerable benefit of comments and suggestions from Jens Christiansen, Paul David, David Mowery, and Stanley Engerman. I am grateful also to the National Science Foundation for financial support.

Notes

1 Karl Marx, *Capital*, Vol.1, Modern Library edition, p.406, n.2. All citations to the first volume of *Capital* are to this edition. The pagination is the same as in the Kerr edition.

2 This view goes back more than 50 years in the professional economics literature. See Alvin Hansen, 'The Technological Interpretation of History,' *Quarterly Journal of Economics*, November 1921, pp. 72–83.

3 Karl Marx, *The Poverty of Philosophy* (Moscow, n.d.), p. 105. The sentences which precede the one quoted above make Marx's meaning perfectly clear and reasonable. 'Social relations are closely bound up with productive forces. In acquiring new productive forces men change their mode of production; and in changing their mode of production, in changing the way of earning their living, they change all their social relations.' Moreover, as Marx points out later, 'The hand-mill presupposes a different division of labor from the steam-mill (ibid., p. 127). Surely one need not be a technological determinist to subscribe to these observations.

4 Karl Marx and Friedrich Engels, *The Manifesto of the Communist Party*, in Karl Marx and Friedrich Engels, *Selected Works*, Vol. 1 (Moscow, 1951), p. 34. See also *Capital*, p. 823; *The Poverty of Philosophy*, pp. 129–33; and Karl Marx, *Grundrisse* (New York, 1973), pp. 505–11.

5 Marx, *Capital*, pp. 337–38, 367.

6 'While simple co-operation leaves the mode of working by the individual for the most part unchanged, manufacture thoroughly revolutionizes it, and seizes labor power by its very roots. It converts the laborer into a crippled monstrosity, by forcing his detail dexterity at the expense of a world of productive capabilities and instincts; just as in the States of La Plata they butcher a whole beast for the sake of his hide or his tallow' (ibid., p. 396).

7 Ibid., p. 353. Note also that Marx's panegyrics on the technological dynamism of capitalism apply *not* to capitalism throughout its history, but only to capitalism as it existed in the century or so before the writing of the *Communist Manifesto*. 'The bourgeoisie, *during its rule of scarce one hundred years*, has created more massive and more colossal productive forces than have all preceding generations together' (Marx and Engels, *Manifesto*, p. 37 , emphasis added).

8 For a forthright statement of technological determinism, see the work of the anthropologist Leslie A. White. According to White, a social system is 'a function of a technological system.' Furthermore, 'Technology is the independent variable, the social system the dependent variable. Social systems are therefore determined by systems of technology; as the latter change, so do the former' (*The Science of Culture* [New York, 1971], p. 365).

9 Marx, *Grundrisse*, p. 278. Emphasis Marx's. Elsewhere he states: 'Whenever a certain stage of maturity has been reached, the specific historical form is discarded and makes way for a higher one. The moment of arrival of such a crisis is disclosed by the depth and breadth attained by the contradictions and antagonisms between the distribution relations, and thus the specific historical form of their corresponding production relations, on the one hand, and the productive forces, the production powers and the development of their agencies, on the other hand. A conflict then ensues between the material development of production and its social form' (Karl Marx, *Capital*, Vol. 3 [Moscow, 1959], p. 861).

10 Marx, *Capital*, Vol. 1, p. 199.

11 Ibid., p. 198.

12 Ibid.

13 Ibid., p. 200, emphasis added. The paleontological mode of reasoning is continued after the passage quoted. Marx adds in a footnote on the same page: 'However little our written histories up to this time notice the development of material production, which is the basis of all social life, and therefore of all real history, yet prehistoric times have been classified in accordance with the results, not of so-called historical, but of materialistic investigations. These periods have been divided, to correspond with the materials from which their implements and weapons are made, viz., into the stone, the bronze, and the iron ages.'

14 Simon Kuznets, *Modern Economic Growth* (New Haven, 1966), p. 9.

15 *Capital*, Vol. 1, pp. 371–72.

16 Ibid., p. 504.

17 Ibid., p. 532. For further discussion of these and related issues, see Nathan Rosenberg, 'Karl Marx on the Economic Role of Science,' *Journal of Political Economy,* July–August 1974, pp. 713–28.

18 For some tentative exploration of these relationships, see Nathan Rosenberg, 'Science, Invention and Economic Growth,' *Economic Journal*, March 1974, pp. 90–108.

19 'As inventions increased in number, and the demand for the newly discovered machines grew larger, the machine-making industry split up, more and more, into numerous independent branches, and division of labor in these manufactures more and more developed. Here, then, we see in Manufacturing the immediate technical

foundation of Modern Industry. Manufacturing produced the machinery, by means of which Modern Industry abolished the handicraft and manufacturing systems in those spheres of production that it first seized upon' (*Capital*, Vol. 1, p. 417).

20 Ibid., p. 418.

21 Ibid., p. 420. See also pp. 417–8 and 421.

22 Ibid., p. 492.

23 Karl Marx, *Capital*, Vol. 3, p. 79.

24 Ibid., p. 100. Later Marx adds: 'The most striking example of utilizing waste is furnished by the chemical industry. It utilizes not only its own waste, for which it finds new uses, but also that of many other industries. For instance, it converts the formerly almost useless gas-tar into aniline dyes, alizarin, and, more recently, even into drugs' (ibid., p. 102).

25 Ibid., p. 80.

26 Ibid., p. 81. See also p. 84.

27 Ibid., pp. 81–82.

28 Ibid., p. 71.

29 'From the moment that the tool is taken from man, and fitted into a mechanism, a machine takes the place of a mere implement' (*Capital*, Vol. 1, p. 408).

30 Ibid.

31 Ibid., p. 420.

32 Ibid., p. 418.

33 See Abbott Payson Usher, *A History of Mechanical Inventions*, rev. ed. (Cambridge, Mass., 1954), especially Chapter 4, 'The Emergence of Novelty in Thought and Action.'

34 In all of this, the natural environment plays a critical role. 'It is not the tropics with their luxuriant vegetation, but the temperate zone that is the mother country of capital. It is not the mere fertility of the soil, but the differentiation of the soil, the variety of its natural products, the changes of the seasons, which form the physical basis for the social division of labor, and which, by changes in the natural surroundings, spur man on to the multiplication of his wants, his capabilities, his means and modes of labor. It is the necessity of bringing a natural force under the control of society, of economizing, of appropriating or subduing it on a large scale by the work of man's hand, that first plays the decisive part in the history of industry' (*Capital*, pp. 563–64).

35 *A History of Mechanical Inventions*, p. 78.

36 Ibid., p. 65.

37 *Capital*, Vol. 3, p. 112.

38 Ibid.

39 Ibid., p. 103, and *Capital*, Vol. 1, p. 442. Marx cites Babbage for supporting evidence in both places and Ure in the former. Both arguments — with respect to improvements in machine design and

improvements in techniques for producing the machines — are combined in *Capital*, Vol. 1, p.442: "When machinery is first introduced into an industry, new methods of reproducing it more cheaply follow blow upon blow, and so do improvements, that not only affect individual parts and details of the machine, but its entire build. It is, therefore, in the early days of the life of machinery that this special incentive to the prolongation of the working day makes itself felt most acutely.'

40 See Nathan Rosenberg, 'Factors Affecting the Diffusion of Technology,' *Explorations in Economic History*, Fall 1972, pp. 3–33.

When Adam Smith wrote his immortal elements of economics, automatic machinery being hardly known, he was properly led to regard the division of labour as the grand principle of manufacturing improvement; and he showed, in the example of pin-making, how each handicraftsman, being thereby enabled to perfect himself by practice in one point, became a quicker and cheaper workman. In each brand of manufacture he saw that some parts were, on that principle, of easy execution like the cutting of pin wires into uniform lengths, and some were comparatively difficult, like the formation and fixation of their heads; and therefore he concluded that to each a workman of appropriate value and cost was naturally assigned. This appropriation forms the very essence of the division of labour, and has been constantly made since the origin of society. The ploughman, with powerful hand and skilful eye, has been always hired at high wages to form the furrow, and the ploughboy at low wages, to lead the team. But what was in Dr. Smith's time a topic of useful illustration, cannot now be used without risk of misleading the public mind as to the right principle of manufacturing industry. In fact, the division, or rather adaptation of labour to the different talents of men, is little thought of in factory employment. On the contrary, wherever a process requires a peculiar dexterity and steadiness of hand, it is withdrawn as soon as possible from the *cunning* workmen, who is prone to irregularities of many kinds, and it is placed in charge of a peculiar mechanism, so self-regulating, that a child may superintend it.

The principle of the factory system then is, to substitute mechanical science for hand skill, and the partition of a process into its essential contituents for the division or graduation of labour among artisans. On the handicraft plan, labour more or less skilled, was usually the most expensive element of production – *Materiam superabat opus*; but on the automatic plan, skilled labour gets progressively superseded, and will, eventually, be replaced by mere overlookers of machines.

By the infirmity of human nature it happens, that the more skilful the workman, the more self-willed and intractable he is apt to become, and, of course, the less fit a component of a mechanical system, in which, by occasional irregularities, he may do great damage to the whole. The grand object therefore of the modern manufacturer is, through the union of capital and science, to reduce the task of his work-people to the exercise of vigilance and dexterity – faculties, when concentred to one process, speedily brought to perfection in the young.

— Andrew Ure, *The Philosophy of Manufactures*, 1835

The Politics of Scientific Conceptualization

NORMAN DIAMOND

Regular readers of *Science for the People* are already familiar with some of the ways in which science is inescapably political. And yet there is a more fundamental, less familiar, intrinsic link between science and politics, the implications of which we have barely begun to discern. *In its most basic aspects, the concepts with which scientists organize data and formulate theories, science is inherently political.* Scientific concepts are not simply asymptotic approaches to underlying truth. They are products of a particular social structure and may in turn either reinforce or challenge the social status quo. Not only the daily practice and social use, but also the content of science would be different in a differently organized society. No one interested in building a more humane society can unquestioningly accept present-day science as if it were a given, unable to be radically different.

Origins of New Concepts

Together with many previous generations, we have grown up hearing a series of apocryphal legends from the history of science, myths which seem to indicate that scientific concepts simply follow from the raw data. Copernicus, so the story goes, came to his new understanding by being a better observer of the heavens, Galileo by comparing the rate of fall of objects dropped off the leaning tower of Pisa. Aristotle wrote that men have 32 teeth and women 28, supporting his notion of different female and male natures, whereas Renaissance scientists actually counted and discovered dental parity. Newton's insights presumably followed from his forced apperception of a falling apple. All of these stories are historically spurious, as is their underlying theme.[1] Science is not purely inductive. As Einstein noted,

> There is no inductive method which could lead to the fundamental concepts of physics . . . We now realize with special clarity, how much in error are those theorists who believe that theory comes inductively from experience.[2]

Philosophers of science and even some science textbooks increasingly recognize that factors extrinsic to science influence the formulation of scientific concepts. I shall argue that these extrinsic factors are primarily social, though of course expressed by individuals, and that, far from detracting from science, they are the factors potentially most under human control. Thus there is the possibility of a science in which scientists can take responsibility for their concepts, as a product of and contributor toward a society which is controlled and intentionally shaped by all the people in it.

Scientists who recognize that concepts do not simply derive from raw data, and even that there may be social influences on the formation of concepts, nevertheless mainly continue to believe that their conclusions are responsive only to the correspondence test — whether or not predicted results are verified by experimentation, whether or not they correspond to external reality. No experiment can be designed, however, to test a proposition outside of a conceptual context or in isolation from all other propositions. Rather all experiments test complex theories with multiple components, many of them simply assumed as commonsensical by the experimenter. There is a large margin of choice in evaluating which component to regard as falsified by any experiment. In the history of science there are many instances of scientists from different historical periods observing the same phenomenon or conducting what would seem to an observer to be the same experiment, but interpreting the results quite differently.[3] Scientists really use two different tests of any hypothesis: one is the correspondence test, the other is whether the hypothesis makes sense in terms of how the scientist is used to interpreting reality as a whole. This latter interpretive framework derives mainly from the scientist's existence in a particular society.

As a way of coping with external reality, all human beings develop an interpretive framework, a world-view, which explains our situation in that external reality to ourselves. In developing a world-view, the most important component of our reality, the major part of what we need to explain, is social. Because each of our situations, our activities and social environment, is similar to that of other individuals and dissimilar to that of yet others, we develop an outlook and responses in common with some other people, defining us as a social group. Briefly and too simply, as our situation changes, as our society changes or as our position in the society changes, our explanation to ourself changes also.[4]

All our ideas, whether in science, politics or music, are conditioned by our world-view. They are thus indirectly shaped by our society and our position in it. We develop or accept ideas as they seem to make sense to us in terms of our general explanatory framework. Life in any particular society thus shapes the range of understandings and approaches in any particular realm of thought. As societies change, as world-views change, new ranges of conceptual possibility are opened in every sphere of thought.

Cosmology and World Order

To compensate for the necessarily condensed and postulatory theoretical presentation, let me explore certain aspects of an example of a large-scale change in scientific conceptualization, the Copernican revolution. As might be expected from the preceding argument, the usual textbook accounts of the history are inadequate and significantly misleading. Prior to Copernicus' time, it seemed fairly obvious to people that the earth was the center of the universe and was stationary. Common sense held that, due to its weight, for the earth to move through space would require a continuous external driving force — something out there to keep shoving us. On the other hand the heavens, lacking substance, revolve of their own nature (or later, by their lack of resistance to an initial impetus). Being heavy, the earth would fall to the center of the universe in any case. Further, for the earth to revolve on its axis at the rate necessary to account for the visible movement of the stars rising in the East and setting in the West, centrifugal force would tear it apart, a consideration inapplicable to the even faster revolutions of the weightless heavens. This older, socially determined common sense was supported by empirical evidence. Contained in a finite universe the outermost limits of which were bounded by Heaven, any movement through space by the earth would have to be manifested in changes in the apparent positions of planets and stars, whereas this parallax was not observable.

The older cosmology with its concentric crystalline spheres and its mathematical hypotheses to enable calculation of planetary positions, its epicycles major and minor, deferents, eccentrics and equants,[5] embodied a series of presuppositions which were becoming less meaningful in the century or more before Copernicus. It premised strict limits on human knowledge and control, a rigid chain of being, and relative unimportance of human concerns. There was a split model of reality. Humans could know *a priori* the necessarily simple and perfect paths of the stars and planets, as distinct from human inability to understand the 'buzzing confusion' on earth, and as distinct from the complex mathematicial devices necessary actually to calculate those heavenly paths. (Ptolemy and his successors explicitly denied that the planets could move epicyclically, etc., as it was necessary to imagine them doing in order to calculate.) The heavens, being of different material than the sub-lunar sphere, were thought to be subject to different laws. Change on earth, including social change, was caused primarily by the movement of the stars rather than human effort. There was a fixed hierarchy of value and authority in the universe, from Heaven, through the crystalline spheres, to Earth and below, a fixed position and code of conduct for each group, with man (my first impulse, to substitute 'people' for 'man,' would clearly be anachronistic) in a crucial intermediate position partaking of both soul and clay.

New World-Views and the Copernican Revolution

As the society changed in fundamental ways, new world-views developed, making possible new perceptions in all fields. The changes occurred first in northern Italy, which is where Copernicus studied. Renaissance society was still hierarchic, of course; however, its former rigidity had been shattered. Wealthy new urban strata had developed with the thirteenth century commercial boom. Growing state centralization and moves by monarchs against the landed feudal nobility created new jobs and possibilities of social mobility for commoners. Medieval criteria of hierarchy were now obscured by competing, though often intermingling, hierarchies, embodying different sets of values. A long economic slump and regression toward sharper class distinctions and obstacles to upward mobility in Copernicus' own lifetime only enhanced the widespread sense of

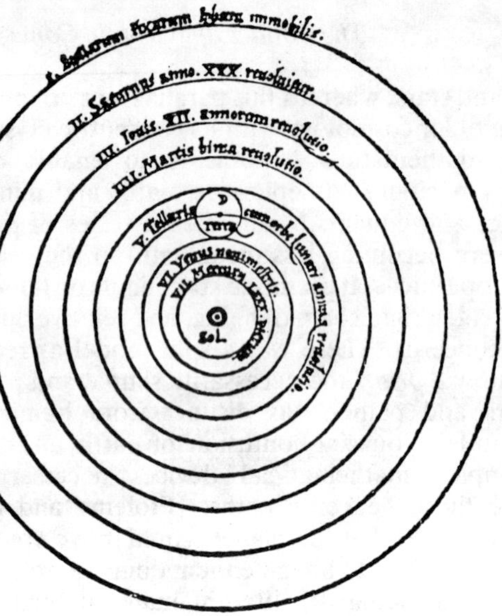

The universe of Copernicus
(from *De Revolutionibus* . . ., 1543)

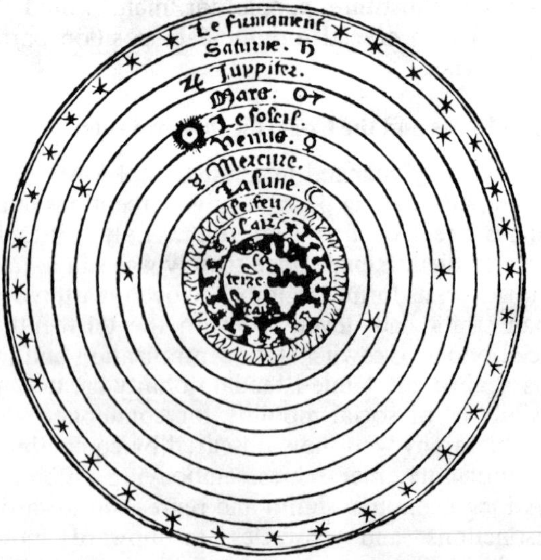

The medieval universe: Ptolemy's celestial spheres
(from O. Finé, *La Theorique des Cielz, Mouvemes, et Termes Practiques des Sept Planetes, nouvellement et tres clerement redigee en langaige fracois*, 1528)

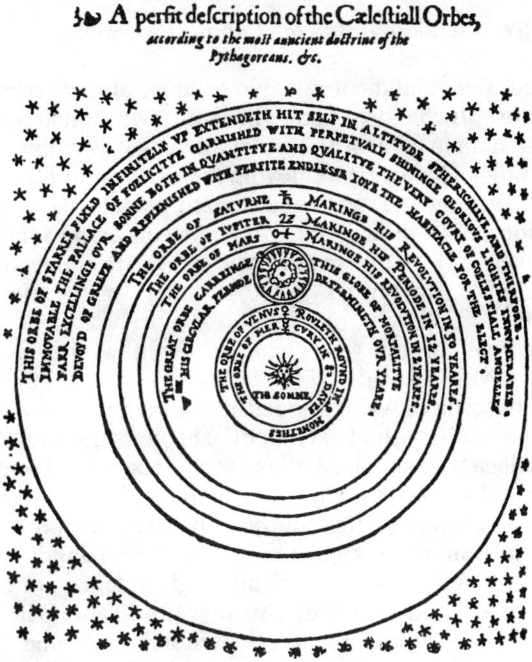

The infinite universe of Thomas Digges
(from *A Perfit Description of the Caelestiall Orbes, according to the most auncient doctrine of the Pythagoreans, latelye revived by Copernicus and by Geometricall Demonstrations Approved* [Based on Book I of Copernicus' *De Revolutionibus Orbium Coelestium*], 1576, folio 43)

'THIS ORBE OF STARRES FIXED INFINITELY VP EXTENDETH HIT SELF IN ALTITUDE SPHERICALLYE, AND THEREFORE IMMOVABLE THE PALLACE OF FOELICITYE GARNISHED WITH PERPETUALL SHININGE GLORIOUS LIGHTES INNUMERABLE, FARR EXCELLINGE OUR SONNE BOTH IN QUANTITYE AND QUALITYE THE VERY COURT OF COELESTIALL ANGELLES DEVOYD OF GREEFE AND REPLENISHED WITH PERFITE ENDLESSE JOYE THE HABITACLE FOR THE ELECT . . . THE GREAT ORBE CARREINGE THIS GLOBE OF MORTALITYE WTH HIS CIRCULAR PERIODE DETERMINETH OUR YEARE.'

In the first English-language version of Copernicus, Thomas Digges depicts a scattering of stars, illustrating his novel concept of infinity. In Copernicus' original diagram, we see a finite outer shell of 'immobile spheres' bounding a heliocentric system composed of mercury, venus, earth, mars, jupiter and saturn. In Finé's diagram of the Ptolemaic system, we see the four elements (earth, water, air, fire) at the centre, with orbits of the moon, mercury, venus, the sun, mars, jupiter, saturn and 'the firmament'.

individual self-fulfillment possible to those already of relatively high status. Much more of one's life and environment was seen to be within human control and understanding. Common sense was undergoing a change. In removing the earth from the center, in rotating it around the sun as one of many planets, the Copernican revolution raised human and mundane concerns to the level of the heavens. It is the manifestation in astronomy of a new interpretative framework corresponding to the new social ambient.

Early adherents were won to the new theory through sympathy with its underlying world-view, not through factors internal to science. As mentioned above, there were no new data or new technology to attract them. The new interpretation was not significantly simpler — in some respects it was more complicated — nor more accurate than the Ptolemaic model (in spite of assertions to the contrary by many textbooks, which distort the content of science as they misunderstand the factors which shape it). It denied the empirical evidence not only that the sun 'rises' and 'sets' but also that observable parallax was absent. Only decades later was the telescope adopted and what was taken to be empirical confirmation provided. Far from being based on new data, Copernicus' theory enabled the perception of new data. Other cultures, not precluded by their socially conditioned general perspectives from seeing celestial change, had observed sunspots and new stars for centuries. After Copernicus, using the same instruments as before, European astronomers began to see these phenomena and to interpret comets as wandering through what before had seemed 'immutable space'. The same premises which enabled the new astronomy and made sense to those of Copernicus' contemporaries whom his argument persuaded, are also to be found in other spheres of thought undergoing decisive transformation at this time. Copernicus' own writing on subjects far from astronomy manifests the same new world-view.

How we organize data in science as in every sphere of consciousness embodies an over-all outlook which derives from our social existence. Underlying and structuring all our thoughts is our understanding of our society and our reactions and adaptations to it. Scientific concepts are thus inherently political, continuing to express and reaffirm socially based world-views. Einstein's reluctance to accept probabilistic quantum theory, to take one modern realization, stemmed explicitly from his rejection of the discordant outlook of which he saw it a product. An excellent recent study (of sexism in the history of biology) in *Science for the People* provides a further illustration of how scientific concepts, in part socially based, in turn reinforce the social status quo.[6] To 'serve the people' with existing science is insufficient.

Practising Politics and Science

For a worker in science who recognizes the need for fundamental social change, the more familiar respects in which science is political lead to relatively limited ways of combining jobs with political activity. Many indeed choose to separate professional from political lives, working with other people after job hours and outside job roles. Others publicize political abuses in connection with science or take advantage of respected positions based on work in science to speak out on social issues. Some scientists or science workers who are radicals organize their co-workers to rearrange or diminish hierarchies in the work situation. And yet all of these approaches leave the science itself, the content of research and formulation of results, untouched. Considered in those terms, science seems to offer fulfillment mainly in ways that are apolitical. For someone who is politically committed, there are constant qualms about whether and how much even to be working on science. Some people become science dropouts to expend energy on efforts more directly political. For others who need to hold a job in science and yet are unable to reconcile science and politics, the tension may result in lessening political commitment.

There is another important political option which derives from the above discussion. It is possible to use one's scientific knowledge to oppose specialization or overcome some of its deleterious effects. Often what pass for narrow technical decisions really contain disguised political decisions which can be extricated and pointed out. *Science for the People* has been full of examples.[7] The aura of technical expertise shelters what are political decisions from question and criticism. There is a political point too in attempting to enhance not only the scientific understanding of non-scientists but also their sense of their own ability to understand. Effective 'popularization' has negative connotations only to people who accept the elitist premises of modern science. And yet, integrating the concepts with which scientists work for presentation to a lay audience, still accepts those concepts as given. *It is through recognizing that scientific concepts themselves are political that it is primarily possible not only to be a radical and a scientist, but to be a radical scientist.*

Every society rests on the consciousness of its members. Their adherence to, or at least acceptance of, its structure is ultimately what holds the society together. The major obstacles we face in doing political organizing in our own society are a widespread lack of ability to conceive of a better society, or more commonly a sense that it is impossible fundamentally to change what we've got. Corporations, the state, etc., all of them clearly opponents, are obstacles primarily because too many people continue to believe them legitimate.

As a large part of this essay has attempted to show, consciousness is not autonomous. There is much about our society that encourages people's cynicism, apathy and low regard for themselves. In political organizing toward social change, it is not possible to work at the level of consciousness alone. There are severe limits on the extent to which people's sense of social alternatives and sense of their own capabilities to help shape those alternatives can be altered without some alteration in their lives. A revolutionary movement, aiming at a society in which all the people will run the society, must engage people increasingly in conscious and active participation toward changing social conditions now. This is distinct from models of revolutionary action which postulate a revolutionary elite as the sole active force, or ones which include an uncomprehending or merely sloganistic mass participation, or ones which would require waiting until after a seizure of power for the entirety of the social change. This is distinct also from any one-way conception of the relationship between social conditions and consciousness, either one that expects capitalism inevitably to fall from its own contradictions, exclusive of human effort, or one that approaches people without careful consideration of the conrete factors that shape and limit their receptivity. The very fact of being part of political struggle is itself a changed social condition which makes possible changes in perspective and attitude, especially if the political struggle is well chosen and well organized.

A crucial objective of organizing is the fundamental transformation of outlook. Through their struggles, people must learn to understand our society, what maintains it and what will be required to change it. There are important answers that will elude us until we have a mass movement with the capacity to shake and test the society. People must see themselves as capable, if united, of effecting basic change and increasingly able themselves to decide which tactics will further our growing knowledge and ability to transform. It is insufficient to be only anti-capitalist, anti-ruling class, anti-racist, etc. More than an abstract idea of the kind of egalitarian, genuinely democratic society toward which we aim is necessary. That society must be seen as a real prospect and legitimate objective. Unfortunately even many radicals deep down do not believe that a better society is anything more than a theoretical possibility. Their actions and the ways they work with others manifest their acceptance of the prevailing order.

Going Beyond: Doing Radical Science

Knowing that science concepts would and will be different in a qualitatively different society[8] enables science workers in their daily practice now to call into question this society and the consciousness that sustains it. This can be done in two basic ways. The first is by learning to identify the hidden, seemingly commonsensical and thus hard to see, premises that underlie accepted concepts, and by learning to recognize how these premises reflect a world-view which is socially based and socially restricted. Showing their connection to the structure of our society, teaching others to understand all ideas and cultural products in social terms, aids people in recognizing that this society is not eternal and cannot be simply accepted as a given.

This first possibility for political practice within science leads to a second. Having discerned the kinds of premises and perspectives promoted by life in this society, radical scientists may begin to be able to develop alternative science concepts based on empathy with a qualitatively better society; to attempt the new possibilities for organizing data which arise out of a different world-view. The difficulties in undertaking this science/political option are formidable, for it requires identifying with a society not yet existent. We are of course fortunate in the availability of socialist societies to present us with alternative models. These can serve, however, to indicate only the barest outlines, the most abstract hopes, for what we could create here. Undoubtedly it is impossible to put oneself entirely outside of one's society. To step back from it, to delegitimate it at its roots within oneself and others is immensely difficult. Yet this is a fundamental goal of radical practice and the preconditon of radical science.

A word is necessary regarding validity in science. Concepts are not arbitrary, nor are they plastic. There is an external reality to which they must correspond. Most of the concepts in present-day science have a definite operational validity (this is not the place to explore other possible tests for validity); they are not, however, the exclusive ways to organize data. Modern science recognizes the tentativeness and incompleteness of any particular concept, the possibility that it will be transformed through further discovery. What is not readily acknowledged is that its supersession, at the level of fundamental conceptual change, is tied to social developments. In addition, scientific concepts are partial not only because they correspond to a particular social structure, as we have seen, but also because most scientists, as a relatively privileged social group, have a stake in only a partial view of their social reality. The world-view which the concepts manifest is thus that of a group barred from an over-all perspective. As partial, the concepts in present-day science have been functional both in controlling natural reality (the operational test) and in not questioning social reality.

To the extent to which it is possible to transcend the dominant world-views based on adaptation to this society, alternative science concepts may be developed now, resulting in a more creative science. This in itself, however, is not a political accomplishment. Without making explicit the political underpinnings, the alternative concepts will become simply creative reinforcements for the status quo. Radical scientists need to be self-conscious of and to show others how their concepts arise, how their ideas relate to their society and how to understand their society. Alternative concepts should be used to raise social alternatives. In science as elsewhere, the theoretical possibility and attainability of a qualitatively better society must be constantly stressed.

There has been a tendency among radicals to reject the usual posing of issues in terms of the scientist's personal or social responsibility, and rightly so.[9] Abuses of science are endemic to an oppressive social order. There is, however, a higher level of individual responsibility which comes with the awareness that concepts do not automatically derive from raw data and are socially influenced. Recognizing that there are choices behind concepts and that these choices have political implications, radical scientists are able to take responsibility for the concepts they use. By doing so they act as precursors of a society in which consciousness is no longer subordinate to social conditions. Through their science now, they can contribute to fundamental social change.

This essay has gone nearly as far as it can. The next step would have to be a start on social analysis and restructuring of current science concepts. This should be read as a challenge to politically committed science workers to undertake that effort.[10]

This example of labour shows strikingly how even the most abstract categories — despite their validity for all epochs (precisely because of their abstractness) — are nevertheless, in the specific character of this abstraction, themselves likewise a product of historic relations, and possess their full validity only for and within these relations . . .

In the succession of the economic categories, as in any other historical, social science, it must not be forgotten that their subject – here, modern bourgeois society – is always what is given, in the head as well as in reality, and that these categories therefore express the forms of being, the characteristics of existence, and often only individual sides of this specific society, this subject and that therefore this society by no means begins only at the point where one can speak of it *as such*; this holds *for science as well*.
— Karl Marx, *Grundrisse*

Notes

1 The data which Copernicus used were in no way significantly superior to Ptolemy's from 1400 years before. As Alexandre Koyré (*The Astronomical Revolution*) points out, the changes in concepts developed almost autonomously from the changes in observational astronomy. Copernicus had access to the more accurate data of Regiomontanus and specifically chose not to use it in his great book, as if to emphasize that what was new was his framework of understanding and not anything empirical. Galileo never carried out the famous experiment at Pisa; if anyone did, it was an opponent. Indeed Galileo remarked that he did experiments only 'to be able to demonstrate to his opponents the truth of his conclusions . . ., though to satisfy his own mind alone he had never felt it necessary to make any.' Aristotle may have miscounted or may have examined only young women before they got their widsdom teeth; however, the difference between his science and that of the Renaissance does not lie in the greater inductiveness of the latter. Aristotle was among the most brilliant of empirical observers. In Galileo's remarks about his predecessors, what he most admired about Copernicus was precisely his ability to deny and get beyond the evidence of his senses. As to Newton, the story is reminiscent of some of the sillier cartoon paraphrases of Mao Tse-Tung's 'Where Do Correct Ideas Come From?', in which a policeman hitting people over the head is seen as the stimulus to revolutionary social knowledge. People have been oppressed for millenia without necessarily revolting; there is no necessary correlation between the quantity or immediacy of oppression and people's ability to see the oppression and recognize the necessity and possibility of alternatives. Quite other factors enter into that recognition, as into the formation of concepts in science.

2 Quoted in Marvin Harris, *The Rise of Anthropological Theory*, p. 288.

3 One example: Methodologically the experiment which Lavoisier performed to discover oxygen was not new. What was new was the conceptual scheme that enabled him to understand the experiment differently than had his predecessors who had used it to confirm the existence of 'dephlogisticated air'. T.S. Kuhn refers to 'the insufficiency of methodological directives, by themselves, to dictate a unique substantive conclusion'; although he would account for the difference in substantive conclusions, mistakenly I think, by factors more internal to science. T.S. Kuhn, *The Structure of Scientific Revolutions*, p. 3.

4 There are many complexities and necessary qualifications which in this short essay must be bypassed. World-views are not usually entirely self-consistent, as the social reality which they explain contains its own contradictory elements. World-views do not

change in simple reflex with changes in the society. Often they incorporate elements of an outlook which corresponded to a previous social situation. The most important respect in which Chinese and Cuban approaches to social transformation differ from the earlier Soviet model is in awareness that full socialist consciousness does not automatically follow changes in the economy and social structure. Both of the later revolutions emphasize intentional changes in people's values and understanding along with engaging people in transforming the social and economic base.

5 Epicycles, etc., are geometric devices which may be indefinitely although cumbrously refined to permit accurate calculations; they allowed the Greeks and their successors to retain the premise that heavenly bodies move in circular orbits. A planet, for instance, would be treated as if its path were that of a point on a rotating circle whose center is moving in a circle around another point.

6 Rita Arditti, 'Women as Objects: Science and Sexual Politics', *Science for the People*, September, 1973.

7 On the choice between nuclear fission and fusion, see 'Energy Crisis', *Science for the People*, January, 1974, especially pp. 12-15. On the 'green revolution,' see K.R. Bhattacharya, 'Soyabeans in India?', *Science for the People*, January, 1974. On priorities in cancer research, see Michele Fluck, 'Cancer Prevention', *Science for the People*, November, 1974.

8 Again, there is no implication that ideas change automatically in one-to-one correspondence with social change. Elements of ideas from previous world-views, from previous social structures, are retained long after the context that gave rise to them or permitted them has been altered. They are retained selectively, however, according to what continues to make sense in terms of people's new social experience and setting. A modern reader of Newton, for example, is struck by the distortion of his intentions and of the interconnection of his thoughts represented in the selective culling his ideas receive in secondary accounts today. It is not so much that each generation rewrites the past as that each social configuration understands the same past differently because it has something different to understand about itself.

9 See, e.g., Hilary Rose and Steven Rose, 'The Radicalisation of Science', *The Socialist Register*, 1972.

10 For the work of a contemporary natural scientist whose socialist commitment enhances his scientific formulations, see Steven Rose, *The Conscious Brain*.

The Taylorisation of Intellectual Work

MIKE COOLEY

The use of fixed capital in the productive process — that is, machinery and latterly computers — marked a fundamental change in the mode of production. It cannot be viewed merely as an increase in the rate at which tools are used to act on raw material. The hand tool was entirely animated by the worker, and the rate at which the commodity was produced — and the quality of it — depended (apart from the raw materials , market forces and supervision) on the strength, tenacity, dexterity and ingenuity of the worker. With fixed capital — that is, the machine — it is quite the contrary, in that the method of work is transformed as regards its use value into that form most suitable for fixed capital. The scientific knowledge which predetermines the speeds and feeds of the machine, and the sequential movements of its inanimate parts, the mathematics used in compiling the numerical control programme, do not exist in the consciousness of the operator; they are external and act through the machine as an alien force.

Although science is merely the accumulation of the knowledge and skill now appropriated, it confronts workers through fixed capital as an alien and hostile force, and further subordinates them to the machine. The nature of their activity, the movements of their limbs, the rate and sequence of those movements — all these are determined in quite minute detail by the 'scientific' requirements of fixed capital. Thus objectivised labour in the form of fixed capital emerges in the productive process as a dominating force opposed to living labour. We shall see subsequently, when we examine concrete situations at the point of production, that fixed capital appropriates not only *living* labour, but in its sophisticated forms (computer hardware and software) appropriates the scientific and intellectual output of the white-collar workers whose *own intellects* oppose them also as an alien force.

Therefore, the more that workers put into the object of their labour, the less there remains of themselves. The welder at General Motors who takes a robotic welding device and guides its probes through the welding procedures of a car body is, on the one hand, building skill into the machine, and on the other, deskilling himself. The accumulation of years of welding experience is absorbed by the robot's self-programming systems and will never be forgotten. Similarly, mathematicians working as stressmen in an aircraft company may design a software package for the stress analysis of airframe structures and suffer the same consequences in their jobs. In each case they have given part of themselves to the machine and in so doing have conferred life on the object of their labour — except that now this life no longer belongs to them but to the owners of the object.

Since the product of their labour does not belong to the workers, but to the owner of the means of production in whose service the work is done and for whose benefit the product of labour is produced, it necessarily follows that the object of the workers' labour confronts them as an alien and hostile force, since it is used in the interests of the owner of the means of production. Thus this 'loss of self' of the worker is but a manifestation of the fundamental contradictions at the economic base of our society. It expresses the antagonistic contradiction between the interest of capital and labour, between the exploiter and the exploited. Therefore at this historical stage fixed capital is the embodiment of a contradiction — namely, that the means which could make possible the liberation of the workers from routine, soul-destroying, backbreaking tasks, are simultaneously the means of their own enslavement. It is therefore obvious that the major contradiction can be resolved only when a change in the ownership of the means of production takes place.

I have attempted to show elsewhere that this loss of self and alienation are now spreading rapidly, in the wake of computerisation, into highly creative fields of intellectual work such as architecture and scientific fields of work such as aerospace design.[1] Technological change of this kind means that the organic composition of capital is changed. Processes become capital-intensive rather than labour-intensive as workers by hand and brain are replaced by 'dead labour', i.e., by very sophisticated

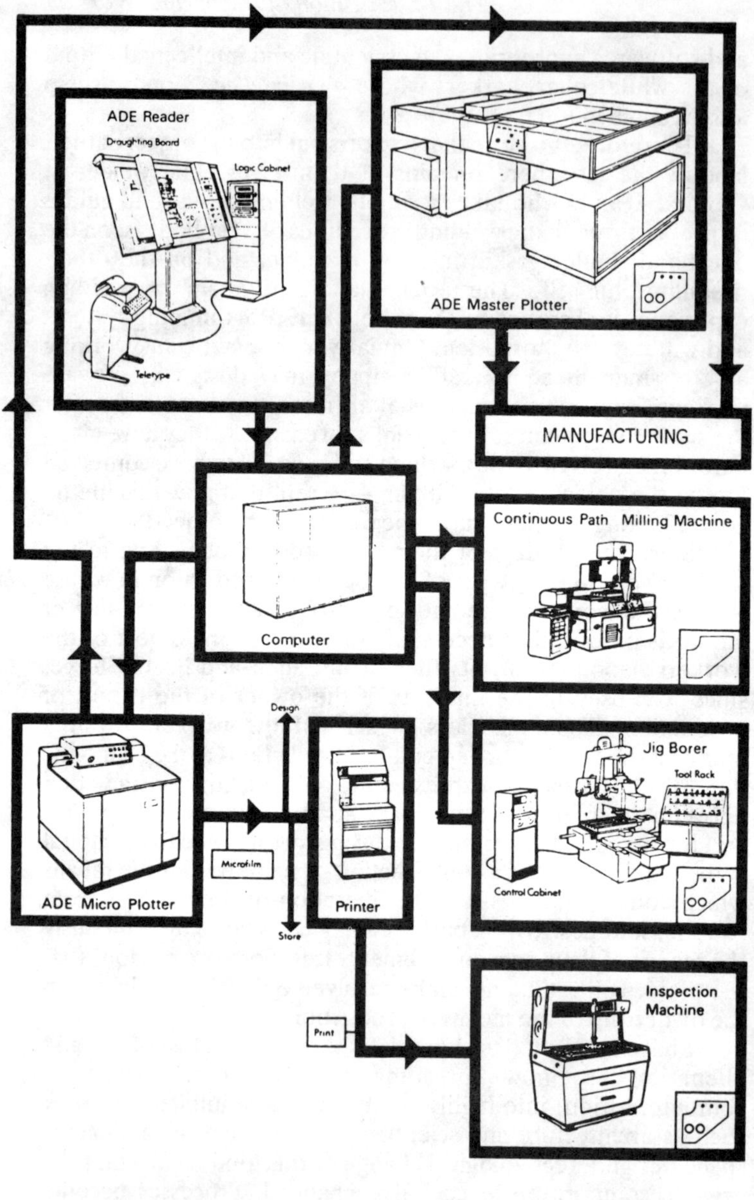

ADE SYSTEM (Automatic Drafting Equipment)
Systems of this kind permanently eliminate some of the most highly skilled work on the shopfloor.

machines. This is now giving rise to massive structural unemployment throughout most of the technologically advanced nations, and is a tendency which is going to continue. It does however mean that groups of 'workers' such as scientists and technologists, who have never previously had experience of the degradation of the dole queue, are now feeling all the pangs of job insecurity[2] and are therefore much more likely to align themselves with those sections of the working class, and critics of society, who recognise the need for fundamental changes in the kind of priorities our society is setting itself.

In addition there is the disillusionment and the loss of job satisfaction which flows from the general tendency to reduce the level of qualification in all jobs.[3] Few manual workers any longer harbour illusions that science and advanced technology will liberate them from soul-destroying, backbreaking routine tasks and leave them free to engage in more creative work. This realisation flows from the sheer grinding day-to-day experience of the dehumanised alienated tasks in modern production. This dehumanisation is due in large part to the fragmentation of work into narrow tasks, each minutely timed — the very essence of Taylorism.

As its founder Frederick Taylor explained, 'The workman is told minutely just what he is to do and how he is to do it, and any improvement he makes upon the orders given to him is fatal to success.'[4] Initially, Taylorism on the shop floor actually increased the intellectual activity of staff in the offices. As Taylor himself explained in his long paper, 'Shop Management' (1903), his system is 'aimed at establishing a clearcut and novel division of mental and manual labour throughout the workshops. It is based upon the precise time and motion study of each worker's job in isolation, and relegates the entire mental parts of the tasks in hand to the managerial staff.'

Seventy years of scientific management have seen the fragmentation of work grind through the spectrum of workshop activity engulfing even the most creative and satisfying jobs such as toolmaking. Throughout that period most industrial laboratories, design offices and administrative centres were the havens of the conceptual, planning and administrative aspects of work. In those areas, one spur to output was the dedication to the task in hand, an interest in it, and the satisfaction of dealing with the job from start to finish. The objective base for Taylorism was laid in the field of intellectual work when large numbers of scientific and technical workers were involved in the labour forces as a whole. It was evident that, eventually, management would divide up their intellectual activity into routine tasks and would apply work study to set precise times for its synchronisation with the rest of the 'rational procedure'.

In the 1830s the father of the computer industry, Charles Babbage, anticipated Taylorism in the intellectual field. He wrote:'We have already mentioned what may perhaps seem paradoxical to some of our readers — that the division of labour can be applied with equal success to mental as well as mechanical operations, and that it ensures in both the same economy of time.'[5] A paper in the journal *Work Study* (June 1974, p.23) entitled 'A Classification and Terminology of Mental Work' suggests that much 'progress' has been made in this direction. Having identified the hierarchy of physical work as 'Job, Operation, Element,Therblig', it states:

> The first three of these are general concepts, i.e. they could apply equally to mental or physical work. The last term, the therblig, is specific to physical work. All elements of physical work consist of a small number of basic physical motions, first codified by Gilbreth (therblig=anagram of Gilbreth), and later amended by the American Society of Mechanical Engineers and in the British Standard Glossary. The logical pattern would be complete if a similar breakdown of elements into basic mental Motions or yalcs were available.

The paper discusses in detail how to classify input yalcs, output yalcs and processing yalcs. It describes how each of these can be subdivided down into basic mental operations. It goes so far as to draw a division between passive reception of visual signals (seeing) and active reception (looking), and passive reception of auditory signals (hearing) and active reception (listening). The paper implies that these techniques will be used in the more clerical aspects of mental work. However it concludes:

> We have tried to show that mental work is a valid and practical field of application for "Work Study", that basic mental motions exist and can be identified and classified in a meaningful way provided that one does not trespass too far into the more complex mental routines and processes. A set of basic mental motions has been identified, named, described and coded as a basis for future measurement research, leading to the compilation of standard times. There is a good prospect that such times would play a valuable part in Work Study Projects.

Whether or not one regards this type of research as

pseudoscientific, there can be little doubt about how it will be deployed. The employers of scientific, technical and administrative staff (including some forms of managerial staff) will see it as a powerful form of psychological intimidation to mould their intellectual workers to the mental production line. It is perhaps a recognition of this tactical importance which prompted Howard C. Carlsson, a General Motors psychologist, to say: 'The Computer may be to middle management what the assembly line is to the hourly worker.' Industrial experience already demonstrates that Taylorism is destined not just for those who use computers or are in middle management, but for a range of intellectual 'workers' including scientists and technologists.

It is true that some intellectual workers have deliberately chosen to work in an academic environment in order to free themselves from the rat race and from the pressures which are typical of modern industry. It is also true, of course, that some of these academics engaged in so-called 'Neutral Science' were quite indifferent to the consequences their work was having upon workers at the point of production. We now have the rather ironic development in which some university researchers — having helped to develop the scientific management production systems which made work so grotesque for those on the shop floor — may themselves soon become victims of their own repressive techniques. An article entitled 'College of Business Administration as a Production System' (*Academy of Management Journal*, Vol. 16, no. 2, p.306) is symptomatic of a general tendency.

The terminology used to describe academic features and activities in the form of a factory model is strongly indicative of the underlying philosophy. Thus the recruitment of students is referred to as 'material procurement', recruiting of faculty as 'resource planning and development', faculty research and study as 'supplies procurement', instructional methods planning as 'process planning', examinations and award of credits as 'Quality Control', instructor evaluation as 'resource maintenance' and graduation as 'delivery'. The professors and the lecturers are of course 'operators' and presumably, as on the shop floor, only the effective operators can be tolerated (effective for what? and for whom? we may ask).

Those academics engaged in the physical and pure sciences will be pleased to learn that the important issues of efficiency and optimisation will not be left to the subjective ramblings of the sociologist or the tainted ideology of the political economist. The full analytical power and neutrality of real science and the penetrating logic of mathematical method have been brought to bear. They will undoubtedly produce a completely 'objective' solution. I have referred elsewhere to the use of the well-known Franke-Wolfe algorithm by Geoffrion, Dyer and Freiberg who define six criteria for optimising efficiency in the university, and then proceed to show how it can be determined.[6]

Programmes to increase faculty productivity are already spreading rapidly in the United States; recent grants indicate the potential scale. For example, a centre for professional development was set up in the California State University and College System with a grant of $341,261 from the Fund for the Improvement of Post-Secondary Education in Washington. A comprehensive faculty activity-analysis questionnaire has been prepared and developed by the University of Washington. The actual percentage time devoted to each faculty activity is requested. Each activity is specified very precisely as it might be in a factory situation. Even one's own reading activities are included. There are some academics who hope that, in projects of this kind, educational requirement will outweigh mere productivity considerations. However, a growing sector feel the outcome will be a smaller faculty teaching more classes and more students. This has already happened at the City University of New York, where 700 faculty members were sacked (*New Scientist*, 22 April 1976, p178).

However, increased 'productivity' could have consequences much more widespread and subtle than the obvious ones of increased work tempo, loss of control, job insecurity and even redundancy. The impact it will have on the creativity of those involved is likely to be significant, for central to all optimisation procedures of this kind is the notion of specific goal objectives. A vivid example of the need to avoid such an over-constrained work environment was the design recently of E.M.I.'s computer-controlled brain and body scanner. In his evidence to the Parliamentary Select Committee on Science and Technology, E.M.I.'s Managing Director Dr John Powell pointed out that the scanner was developed using un-allocated funds as a by-product of work on optical character recognition. He stated in evidence that had its inventor 'been constrained to follow a set objective on contract research funded by an operating division he might have just produced another optical character recognition machine'.

Thus even the elitist right of the scientific worker or the researcher to give vent to his or her creativity will now be increasingly curbed by the system as it seeks more and more to control human behaviour in all its aspects. This is part of the general process of the small elite who control society to gain complete control over all those who work, whether by hand or by brain, and who use scientific management and notions of efficiency as a vehicle for doing so. It will be seen, then, that the organisation of work, the means of designing both jobs and the machines and computers necessary to perform them, all embody a definite choice of values. So, by regarding science and technology as neutral, we have

> failed to recognise as anti-human, and consequently to oppose the effects of, values built into the apparatus, instruments and machines of their capitalist technological system. So, machines have played the part of a Trojan Horse in their relation to the Labour Movement. Productivity becomes more important than fraternity. Discipline outweighs freedom. The product is in fact more important than the producer, even in countries struggling for socialism.[7]

It has been suggested that by ignoring these considerations the Soviet Union was laying the basis for the present situation, in which it would be hard to argue that a worker enjoys a sense of fulfilment through his or her work as envisaged by the early Marxists. It may well be that, by trying to adapt the forms of science and technology developed in the capitalist countries to their own country, the Soviet Union foreclosed the possibility of developing entirely different forms. Indeed the development in that country must find part of its origins in the attitude Lenin took to Taylorism, of which he said:

> Like all capitalist progress [it] is a combination of the refined brutality of bourgeois exploitation, and a number of the greatest scientific achievements in the field of analysing mechanical motions during work, the elimination of superfluous and awkward motions, the elaboration of the correct methods of work, the introduction of the best system of accounting and control etc. The Soviet Republic must at all costs adapt all that is valuable in the achievement of Science and Technology in this field. The possibility of building socialism depends exactly on our success in combining the Soviet Power and the Soviet Organisation of Industry with the up-to-date achievements of capitalism. We must organise in Russia the study and teaching of the Taylor system, and systematically try it out and adapt it to our ends.[9]

Socialism, if it is to mean anything, must mean more freedom rather than less. If a worker is constrained through Taylorism at the point of production, it is inconceivable that he or she will develop the self-confidence and the range of skills, abilities and talents which will make it possible to play a vigorous and creative part in society as a whole. So it would appear that, in the technologically advanced nations, there are now beginning to emerge a range of contradictions which will necessitate a radical examination of how we use science and technology, and how knowledge should be applied in society to extend freedom and development.

It is precisely problems of this kind — contradictions experienced in the course of their own work — which have caused more and more discussion about these issues among workers: firstly, in France during the late 60s, and currently in the United Kingdom. Issues of this kind were central to the development in Lucas Aerospace of the Workers' Corporate Plan. There, scientists and technologists realised that they had a direct community of interest with those on the shop floor, and they established a Combine Committee which, in the words of its Secretary Ernie Scarbrow, 'combined on the one hand the analytical power of the scientist and technologist, with the equally (or more) important common sense of those on the shop floor on the other'. This, he pointed out, provided a unique and volatile combination. It is the kind of combination that one hopes will be witnessed in more and more large plants throughout the United Kingdom. In that Company, the highest-level technologists and workers on the shop floor were together subjected to the degradation of structural unemployment. But not only that; they saw also the gap between that which their ingenuity, skill and ability could provide, and that which it was actually providing in our society. They were engaged during the day in designing, developing and building equipment to be used in an aerospace system as sophisticated as Concorde, whilst some of them were passing through communities (or living in them) where old-age pensioners were dying of hypothermia. They were also subjected to the gradual deskilling of their functions, and a complete loss of control. This deskilling, as in other large multinationals, denied them the right of any freedom of expression in the course of their daily work.

So they proposed (in their Corporate Plan of six volumes) a whole range of products which society needs, and which they had the skills and abilities to produce.[10] They questioned the underlying assumptions of the Market Economy which prevented them doing that, and began to consider products for their use value rather than their exchange value. They also opened a debate in which, as workers by hand and brain, they thought of themselves in their dual role in society — that is, both as producers and as consumers. Lastly, and probably most importantly, they proposed an entire re-organisation of the labour process, in which hand and brain could again be linked, and workers could give full vent to their creativity. For the Lucas workers believe 'that Society's greatest asset is the skill, the ability, the ingenuity and the energy of its ordinary people'.

Although the social and political implications of the Lucas Aerospace Workers' Plan have been widely discussed, the technology contained within it has been largely ignored. This is particularly true of criticisms of the plan.[11] This reluctance to deal with the technology of the Plan is on the one hand due to the left's remarkable ignorance of science and technology, and on the other an indifference to it, because it is perceived as 'neutral'.

Yet for those at the point of production, the considerations of the technology, the design methodologies, and the nature of the labour process which arises from them are all of equal importance to the political considerations, precisely because these workers do not separate one from the other. Indeed, one of the most positive features the Lucas workers see arising from their Corporate Plan are the discussions now taking place with Shop Stewards and representatives of workers at all levels in industry — from scientists to semi-skilled workers, in a range of companies spanning Vickers, Parsons, Rolls Royce, Chrysler, Dunlop and Thorn. These dicussions not only take up the conventionally 'political' aspects but are giving rise to a profound questioning of the nature of the technology itself and the design methodologies used.

The Lucas workers are deeply conscious that, if the debate were limited to industrial workers of their own kind, it would represent a new form of elitism. They are therefore making strenuous efforts to involve wide sections of the community at large in such discussions, so that they can interact with them and learn from them. To this end, they are working with the North East London Polytechnic to convert a coach into one of their hybrid road-rail vehicles.[12] They hope this vehicle will be a form of technological agitprop. It is intended to have an exhibition

THE LUCAS *AEROSPACE WORKERS*
ROAD RAIL VEHICLE

Poster ©Cliff Harper

The Road-Rail Mechanism is shown here removed from the front end of the bus. It is special because its pneumatic tyres run on both the roads and the rail. When the vehicle runs on the rails the flanged guide wheels take only a small proportion of the load and actually steer the main pneumatic tyres along the track. When the vehicle approaches a curve in the track the guide axle follows the curve and hence pushes the lever link to one side. The lever movement here is transfered via the steering links to slightly turn all of the four pneumatic tyres to let the vehicle safely negotiate the curve.

The vehicle is able to switch smoothly between road and rail using the retractable guide wheels. A second guide axle at the back of the bus ensures that the rear is held on the track.

The Lucas Workers are skilled at designing and making electronic and mechanical control equipment for such aircraft as Concord and the Tornado. These skills could easily be used to produce lightweight, safe and reliable Road-Rail Vehicles.

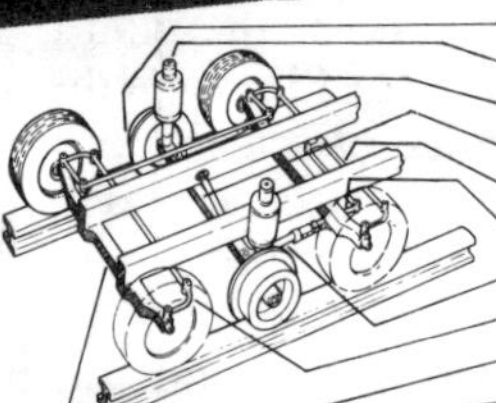

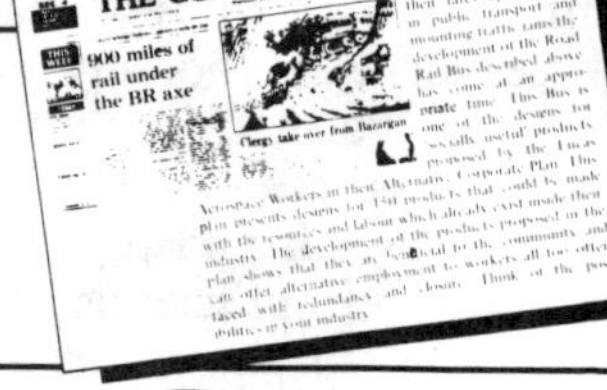

With British Rail putting their fares up, cutbacks in public transport and mounting traffic jams the development of the Road Rail Bus described above has come at an appropriate time. This Bus is one of the designs for socially useful products proposed by the Lucas Aerospace Workers in their Alternative Corporate Plan. This plan presents designs for 150 products that could be made with the resources and labour which already exist inside their industry. The development of the products proposed in the plan shows that they are beneficial to the community and can offer alternative employment to workers all too often faced with redundancy and closure. Think of the possibilities in your industry.

The Road-Rail Bus is shown opposite serving local communities. Passengers can be picked up from outlying villages not served by rail and driven to the nearest rail track. They would then continue at much higher speeds and greater comfort along the branch lines, collecting other passengers on the way. When the bus arrives at the larger towns and cities it can drive its' passengers, by road, to their final destinations.

The Road-Rail idea is not confined to this country - it has its uses in third world countries where little or no rail network exists. The Truck-Train pictured here uses the same principle as the Road Rail Bus. Instead of running on narrow steel rails a special concrete track is laid. The steering mechanism picks up on the inside edge of the track. Such a Truck Train could carry heavy goods loads over long distances very economically. The rubber tyres grip the track more firmly than ordinary steel railway wheels. This means that the vehicle can climb much steeper gradients, hence reducing the cost of laying track. An added bonus is that each goods wagon can be used to deliver the load by road to the final destination - making it a truly integrated transport system.

SUPPORT THE RAIL - BUS FUND

DONATIONS TO C.A.I.T.S. N.E.L.P., LONGBRIDGE RD., DAGENHAM, ESSEX

within the coach, with photographs, slides and videotapes describing the concepts underlying the Corporate Plan,[13] and showing some of the prototypes of the products in action. It is hoped that local trade union branches and trades councils, together with community groups, will sponsor the visits of the vehicle to the different cities. These visits could culminate in large public meetings where 'ordinary' people can begin to discuss with technologists and industrial workers what type of products they would really like to have.

In the meantime, they continue to debate, in considerable detail, those forms of technology which will give full vent to the creativity of the hands and minds of the workers, and could be produced by non-hierarchical forms of industrial organisation. In the course of actually designing and building prototypes, they have discovered, as one of them expressed it, '. . . that management is not a skill or craft or a profession but a command relationship, a sort of bad habit which we have inherited from the army and the church'. They have also shown, if only in embryo, that the design methodology used in a 'socialist technology' would have to be radically different from that which applies in our current technology.

At present, in the technologically advanced nations, highly qualified designers and technologists spend months drawing, stressing and analysing a prototype before telling the workers on the shop floor what should be done. These design stages involve rarified, complex mathematical procedures which are necessary mainly because, for commercial reasons, materials have to be exploited to the full. Both the materials and the systems of the products are designed just to perform a precisely defined function for a very short length of time before the product is rendered redundant (planned obsolescence). These rarified mathematical procedures are outside the experience of the mass of industrial workers and are used as a means of bludgeoning their common sense into silence. All their knowledge of the physical world about them, acquired through years of making things and seeing them break and rupture, is regarded as insignificant, irrelevant or even dangerous.

However, with the prototypes developed from the products proposed in the Corporate Plan, the methodology was quite the reverse. Workers on the shop floor had every opportunity to give full vent to their skills and creativity because the protoypes were developed more by 'feel' than by 'analysis'. There is nothing mysterious about this feel or 'tacit knowledge', as I will call it. It comes about as a result of years of direct experience at the point of production itself. It is a sad reflection on the specific form technology takes in this society, that this wealth of knowledge is deliberately eliminated. Indeed, this elimination is the very root of so-called scientific management or Taylorism. Clearly, any talk of industrial democracy within this kind of technology is simply a deception.

It will undoubtedly be argued, by the authoritarians both of the right and of the left, that the Lucas workers' approach to technology is romantic, unrigorous and unscientific. Such a view ignores the fact that a desire to meet real social needs is a vitally important stimulus to good quality, creative design, and that it is a qualitative element of design which cannot be treated in a mathematical fashion as can the quantitative elements.

Nor are the Lucas workers alone in taking this viewpoint of science and technology. In a recent paper, one of this country's leading technologists and academics, Howard Rosenbrock, had this to say:

> My own conclusion is that engineering is an art rather than a science and by saying this I imply a higher, not a lower status. Scientific knowledge and mathematical analysis enter into engineering in an indispensible way and their role will continually increase. But engineering also contains elements of experience and judgement as regards all social considerations and the most effective ways of using human labour. These partly embody knowledge which has not yet been reduced to exact and mathematical form. They also embody value judgements which are not amenable to the scientific method.[14]

That this approach to technology is not only more democratic, but also much more fruitful, is demonstrated by the impressive range of products which the Lucas workers have proposed, and in some instances are building and testing in prototype. One of the most important political and technological proposals in the Corporate Plan is for the design of 'telechiric' (hands at a distance) devices. With these systems, the human being would be in control, 'realtime', all the time; the system would merely mimic human activity, but not objectivise it. Thus the producer would dominate production, and the skill and ingenuity of the worker by hand and brain would be central to the activity and would continue to grow and develop. By linking human intelligence with advanced technology, these devices would help to reverse the historical tendency to objectivise human knowledge and thereby confront the worker with an alien and opposite force (as described earlier).

The methods by which the Lucas workers arrived at the concept of this product range, are in themselves revealing. It was suggested to them that it would be highly socially responsible if a means could be found of protecting maintenance workers on North Sea oil pipelines. These maintenance workers experience a very high accident rate because of the depths at which they have to work.

Since they had been conditioned by traditional design methods, they immediately thought of a robotic device which would eliminate the human being completely. However, the more they began to consider the programming problems of getting a system which would recognise which way a hexagon nut was about (much less if it had a barnacle on it!), and select the correct spanner and apply the correct torque, the more they began to recognise what a difficult task this would be. Yet it is the kind of task which skilled workers can perform even 'without thinking about it'. They simply have to look at the diameter of a nut and bolt, and will know through years of experience what torque they can apply to it without wringing it off and yet at the same time tighten it sufficiently so that it won't come loose again. This they can do without any 'scientific knowledge' such as torsional rigidity of the bolt or the shear strength of its material, yet they will get it right repeatedly. As Polanyi once put it, 'There are things we know but cannot tell.' That is to say, workers don't express this knowledge in its written and spoken form. They demonstrate their knowledge and intelligence by what they do!

Indeed, the Lucas workers began to realise that in Western society we are confusing linguistic ability with intelligence. They then began to compare the levels of intelligence of robotic equipment of that kind with total human information processing capability. While comparisons of that kind are notoriously difficult, the order of things is the machine 10^3 to 10^4 and the human being 10^{14}. However, this 10^{14} brings with it consciousness, will, imagination, ideology, political aspirations; and these are precisely the attributes which employers regard as disruptive.

One doesn't have to be a sociological Einstein to work this out. The multinationals and the employers are so arrogant that they actually put it in writing in case we might fail to understand. Thus the *Engineer* had a headline recently which said, 'People are trouble, but machines obey.'[15] It is therefore no accident that systems are designed around the trivial 10^3 whereas the 10^{14} is deliberately suppressed. It is a political decision geared to the power relationships at the point of production. The Lucas workers feel that there are hazardous, dangerous jobs which should be automated out of existence; what they are questioning is the politics of elevating these design methods to universal principles.

It has been pointed out that technological change, viewed thus, has more to do with the exercise of control over the workforce than it has to do with increasing productivity.[16] Indeed Andrew Ure, in his *Philosophy of Manufactures*, put it even more clearly:

> [the industrialists aim to stop any] process which requires peculiar dexterity and steadiness of hand from the cunning workman, and put it in the charge of a mechanism so self-regulating, that a child may superintend it. The grand object therefore of the manufacturer is, through the union of capital and science, to reduce the task of his workpeople to the exercise of vigilance and dexterity [appropriate to a child].[17]

This historical process of deskilling[18] is an important means through which the employer extends his control over his employees. But in a wider sense, it destroys the social and cultural values which surround the exercise of those skills, as well as the means by which they are acquired. Indeed, we seem to have seriously underestimated the educational, cultural and other significance of skill and craftsmanship.[19] What we are now witnessing is an advanced stage of this deskilling process, for even those who are said to exercise the highest level of skills, such as programmers, are themselves deskilled.[20]

Scientific Management is used to reduce the human being to a blind unthinking appendage to the machine, and thus we arrive at a total systems approach in which the human being is integrated into the system and is viewed as an operating unit. This was clearly spelled out by Boguslaw:

> Our immediate concern, let us remember, is the exploitation of the operating unit approach to systems design no matter what materials are used. We must take care to prevent this discussion from degenerating into a single-sided analysis of the complex characteristics of one type of systems material: namely human beings. What we need is an inventory of the way in which human behaviour can be controlled and a description of some instruments which will help us to achieve control. If this provides us with

> sufficient "handles" on human materials so that we can think of them as one thinks of metal parts, electrical power or chemical reactions, then we have succeeded in placing human material on the same footing as any other material and can proceed with our problems of systems design.

This, then, is the objective dehumanisation which takes place side by side with this advanced technology. Boguslaw continued:

> There are, however, many disadvantages in the use of human operating units. They are somewhat fragile; they are subject to fatigue, obsolescence, disease and death; they are frequently stupid, unreliable and limited in memory capacity. But beyond all this, they sometimes seek to design their own circuitry. This in a material is unforgivable. Any system utilising them must devise appropriate safeguards.

Yet it is precisely this ability — to design one's own circuitry, to think for oneself — that distinguishes us from animal forms, and prevents us being reduced to a type of bee-like behaviour.

> In the construction of its cells, a bee puts to shame many an architect. But what distinguishes the worst of architects from the best of bees is namely this: the architect constructs in his imagination that which he will ultimately erect in reality. At the end of every labour process we get a result that already existed in the imagination of the labourer at its commencement.[21]

Thus we see the significance of raising these issues through the very specific proposals surrounding the telechiric devices. The Lucas workers are indeed developing quite profound political ideas. So also are those who, in the wider sense, are proposing human-centred systems, even in the fields of high-level intellectual activity such as design.

Howard Rosenbrock has developed advanced interactive graphic systems which actually place the designer, and human intelligence, at the centre of the design process. He has described the basic mathematical techniques of these systems.[22] Rosenbrock has developed graphic displays from which the designer can assess stability, speed of response, sensitivity to disturbance and other properties of the system.[23] This he and his colleagues did by using the inverse Nyquist Array. Having demonstrated through his computer-aided design system that different value systems can be applied to problems of this kind, Rosenbrock then raises the much wider question as to whether we are not cutting off options in other fields of intellectual work in rather the same way as we have done at an earlier historical stage in the field of manual work. Other computer scientists, such as Weizenbaum, are now seriously questioning where their work is taking society, and its impact upon human beings and their self-image.[24]

What is often lacking, in honest expressions of concern of this kind, is an economic and political analysis of the forces of society which direct science and technology to fulfill specific class roles. Thus the discussions at Lucas and elsewhere should really be viewed in a much wider context of the overall questioning of the way science and technology are developing under advanced capitalist society. It is linked with the wider challenges workers are attempting to make against the way technology is being organised: the Green Bans Movement in Australia; the attempts by Fiat workers in Italy to transcend the narrow economism which characterised trade union activity for so long;[25] and the courageous struggle by the women at Algots Nord in Sweden.[26]

All of these forces can be linked together in a challenge to the system as a whole, and as a forerunner to a transformation of society which will take it away from its present exploitative, hierarchical form to a new type of society, which the founder of cybernetics once said

> differs from those propounded by many fascist successful businessmen and politicians. People of this type prefer an organisation where all information emanates from the top and where there is no feedback. The subordinates are degraded to become effectors of an alleged higher organism. It is easier to set in motion a galley or factory in which human beings are used only to a minor part of their full capacity, rather than create a world in which these human beings may fully develop. Those striving for power believe that a mechanised concept of human beings constitutes a simple way of realising their aspirations to power. It maintains that this easy way to power destroys not only all ethical values in human beings but even our slight aspirations for the continued existence of mankind.[27]

The new technologies highlight the fact that we are at a unique historical turning point. We must not allow our common sense to be bludgeoned into silence by technocratic and scientific jargon, nor should we be intimidated by the determinism of science and technology into believing that the future is already fixed. The future is not 'out there' in the sense that America was out there before Columbus went to 'discover' it. It has still got to be built by human beings and we do have real choices. But these choices will have to be fought for, and the issues are both technical and political at the same time. If we permit capital to represent its political choices as purely technical ones, we may find that 'All our inventions and progress seem to result in endowing material forces with intellectual life, and in stultifying life into a material force.'

We have to decide whether we will fight for our right to be the architects of the future, or allow a tiny minority to reduce us to bee-like responses.

> This paper has been written:
>
> *First.* To point out, through a series of simple illustrations, the great loss which the whole country is suffering through inefficiency in almost all of our daily acts.
>
> *Second.* To try to convince the reader that the remedy for this inefficiency lies in systematic management, rather than in searching for some unusual or extraordinary man.
>
> *Third.* To prove that the best management is a true science, resting upon clearly defined laws, rules, and principles, as a foundation. And further to show that the fundamental principles of scientific management are applicable to all kinds of human activities, from our simplest individual acts to the work of our great corporations, which call for the most elaborate cooperation. And, briefly, through a series of illustrations, to convince the reader that whenever these principles are correctly applied, results must follow which are truly astounding.
>
> This paper was originally prepared for presentation to The American Society of Mechanical Engineers. The illustrations chosen are such as, it is believed, will especially appeal to engineers and to managers of industrial and manufacturing establishments, and also quite as much to all of the men who are working in these establishments. It is hoped, however, that it will be clear to other readers that the same principles can be applied with equal force to all social activities: to the management of our homes; the management of our farms; the management of the business of our tradesmen, large and small; of our churches, our philanthropic institutions, our universities, and our governmental departments.
>
> — Frederick Taylor, *The Principles of Scientific Management*, 1911

Notes

1 Mike Cooley, 'Contradictions of Science and Technology', in H and S Rose (eds.), *The Political Economy of Science*, Macmillan, 1976.
2 Mike Cooley, 'Taylor in the Office', in R. Ottoway (ed.), *Humanising the Workplace*, Croom Helm, 1977.
3 Harry Braverman, *Labor and Monopoly Capital*, NY, Monthly Review Press, 1974.
4 Frederick Taylor, *The Art of Cutting Metals*, 3rd edition revised, NY, ASME, 1906.
5 Charles Babbage, *On the Economy of Machinery and Manufactures* (1833), excerpted in Maxine Berg (ed.), *Technology and Toil in 19th Century Britain*, CSE Books, 1979, pp. 41–55.
6 Mike Cooley, 'The University as Factory', *New Scientist*, *70*, 1006 (24 June 1976).
7 Robert Jungk, *Qualität des Lebens*, Cologne, Eva, 1973.
8 Harry Braverman, op. cit. (note 3).
9 V.I. Lenin, 'The Immediate Tasks of the Soviet Government' (1918), in *Collected Works*, *27*, Moscow, 1965.
10 Mike Cooley, 'Lucas' Socially Useful Prototypes', *Undercurrents* 26 (Feb-Mar 1978), 28-31.
11 Dave Albury, 'Alternative Plans and Revolutionary Strategy', *International Socialism* 6 (Autumn 1979).
12 Richard Fletcher, 'Guided Transport Systems', North East London Polytechnic, 1978.
13 Mike Cooley, 'Design, Technology and Production for Social Needs', *New Universities Quarterly*, Winter 1977.
14 Howard Rosenbrock, 'The Future of Control', *Automatica*, *13* (1977), 389–392.
15 *The Engineer*, 14 September 1978, 24–25.
16 S. Marglin, 'What Do Bosses Do?', in André Gorz (ed.), *The Division of Labour*, Sussex, Harvester, 1976.
17 Andrew Ure, *The Philosophy of Manufactures* (1835), excerpted in Berg, op. cit. (note 5), pp. 65–70.
18 Harry Braverman, op. cit. (note 3).
19 Arthur Clegg, 'Craftsmen and the Origin of Science', *Science & Society*, *43*, 2 (Summer 1979), 186–201.
20 Robert Boguslaw, quoted in *The Political Economy of Science*, op. cit. (note 1), p. 88; see also Philip Kraft, *Programmers and Managers*, NY, Springer Verlag, 1977.
21 Karl Marx, *Capital*, Volume 1, Lawrence & Wishart, 1974, p. 174.
22 Howard Rosenbrock, *Computer Aided Control System Design*, Academic Press, 1974.
23 Howard Rosenbrock, *Interactive Computing: A New Opportunity*, Control Systems Centre Report no. 388, UMIST, September 1977; also Howard Rosenbrock, op. cit. (note 14).

24 Josef Weizenbaum, 'On the Impact of the Computer on Society — How Does One Insult a Machine?', *Science, 176* (1972), 609–614; Josef Weizenbaum, *Computer Power and Human Reason*, San Francisco, Freeman & Co., 1976.

25 *Italy 1977–78: Living With an Earthquake*, Red Notes, 1978; *Working Class Autonomy and the Crisis*, Red Notes & CSE Books, 1979.

26 Cooley, Friberg & Sjöberg, *Alternativ Produktion*, Stockholm, LiberFörlag, 1978.
Algots Nord is a clothing factory which went bankrupt in 1976 and called in a consultancy firm to recommend the most 'painless' form of redundancies, which were announced two years later. The women who comprised the workforce soon recognised the limitations of having to defend their jobs in terms of making the firm profitable. They instead kept the factory running themselves by producing socially useful clothing — including special designs for 'unprofitable' consumers such as the elderly and handicapped — as well as providing repair service. The women challenged both the capitalist labour process and its products by opposing the normal fragmentation of tasks and workers' permanent allocation to a particular task.

27 Norbert Weiner, *The Human Use of Human Beings*, p 16.

Life as a Productive Force:
Capitalising the Science and Technology of Molecular Biology

EDWARD YOXEN

THE FRAMING OF BIOTECHNOLOGY

In January 1980 the prestigious weekly scientific journal *Nature* devoted a dozen of its opening 'science news' pages to the burgeoning field of biotechnology. It was attempting to tell its scientific, industrial and governmental readership just how important — technologically, financially and politically — applied molecular biology has become.[1] Only two weeks later *Nature* returned to the same theme, on the note of militant enthusiasm for massive government subsidy to develop an internationally competitive biotechnology sector in British or European industry.[2] The reason for these vigorous comments on the second occasion was the announcement that scientists working for a recently formed research corporation, Biogen, had succeeded in reprogramming bacteria by using recombinant DNA techniques,[3] so as to produce a substance known as interferon.[4] This dramatic achievement has important commercial implications since interferons, normally extremely expensive to produce, are held to constitute an enormous new market in the medical field, and interferon is just one substance that can now be produced by reprogrammed bacteria.[5] Biogen is but one of about a dozen new companies that have been formed by or have recruited the world's leading molecular biologists to break into this technological race, working with hundreds of millions of dollars of venture capital from large corporations.[6] In short, the tempo of this area of biology is accelerating fast, its industrial appeal is now prodigious, and its political and economic importance is being rapidly transformed. The British government, so *Nature* argued, must make every effort to assist

British scientists and British firms in this race or face technological eclipse and continuing economic decline.[7] What then lies behind these desperate enthusiasms, with multi-million dollar support?

Clearly this is not merely excited rhetoric. There is plenty of evidence that major strategic choices are being made in biological research[8] and in new industrially oriented labs being set up, as molecular biologists react to and lobby for their developing orchestration and domination by capital.[9] Even those who dislike the idea of industrial sponsorship and control are being forced to swim, more or less enthusiastically, with the tide. Admittedly even *Nature* qualified its editorial enthusiasm for biotechnology by including a discussion of possible threats to academic freedom, with the very suggestive analogy of military research on the university campus.[10]

Such a major development in the role of biology as a productive force requires serious analysis so as to give sense and order to an oppositional strategy.[11] I want to contribute to that task by considering the history of molecular biology as a research programme and as a technological project—formed, organised and regulated by economic and political forces. By following through the historical development of this area of science, I want to present recent actions as an attempt to reproduce, sustain and extend capitalist hegemony in and through science, as an attempt to elaborate a system of control in the form of new biotechnologies. In this light the development of molecular biology may be viewed as a linked set of plans, institutional innovations and strategically conceived professional moves, all of which have helped to build up a system of control of biology by capital, without the farsighted prediction or prevision of that as an eventual goal. The process has been one of unequal competition among rival, partial strategies, each of a necessarily limited vision, for developing biology in various ways within a changing system of economic and political power. Its present result is that the industrial exploitation of recombinant DNA research by corporate capital can be serviced directly from the academic research laboratory.

To clarify the sense of my analysis, the discussion continues with three sub-sections which introduce some of the themes and concepts. Subsequent sections go on to make more specific comments about molecular biology and the history of its concerns.

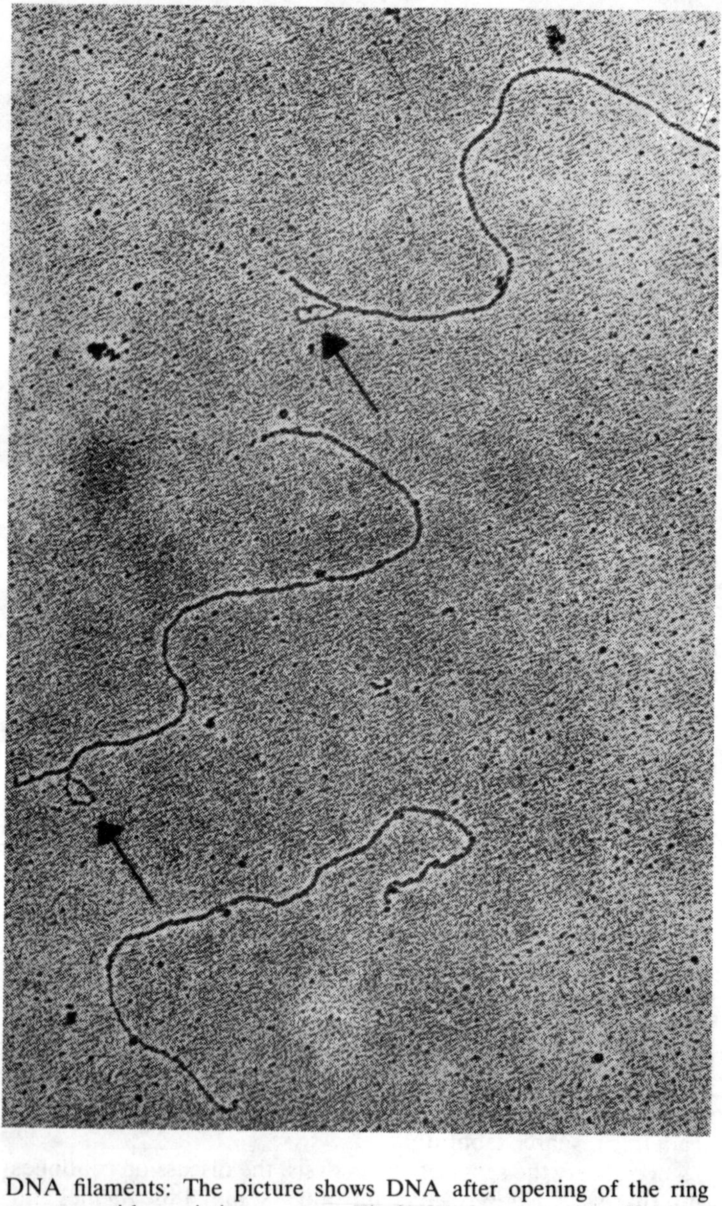

DNA filaments: The picture shows DNA after opening of the ring structure with restriction enzymes. The inserted genes are identifiable as bow-shaped (see arrows). Each of the three DNA filaments are approx. 1/2000 mm long.

Conceptions of Life as Cultural Production

Molecular biology is not only about new institutions, new techniques and new styles of research. It has also allowed the formation of new conceptions of life, nature and humanity, which are fundamentally important to both the pursuit of the research itself and its relation to contemporary society.[12] At the centre of molecular biology in its developed state lies the principle that genes (which in concrete form are sections of double-helical DNA molecules located within more complex structures in cell nuclei known as chromosomes) can be thought as instructions. These instructions are encoded at the level of molecular structure, and together constitute an interrelated, organised set, which is often described in terms of the useful but inexact analogy of a computer program (denoted in this article by the use of the American spelling). This inherited program, the result of the recombination of partial sets of instructions from the preceding generation, underlies and directs the feedback-controlled biosynthetic processes through which organisms are built up and maintained in a stable form. This set of ideas is considered applicable to all organisms, even though important parts of the scheme are missing, such as an explanation of how cells differentiate in development. One aspect of this universality is the notion that the genetic code is the same in all organisms.

Although there is disagreement about how to use the label 'molecular biology', I want to suggest that the distinctive theoretical result of biology at the molecular level over the last 30 years has been the formation and mobilisation of a complex of ideas, redolent of the fields of information theory, cryptography, systems theory and cybernetics—namely program, structure, feedback and code. Indeed molecular biologists often invoke the concept of 'information' in their continuing demarcation dispute with biochemists as the *differentia specifica* of the molecular biological approach, the touchstone or organising principle that raises the subject above the level of mere curiosity about the metabolic transformations of matter and energy. The molecular biological schema purports to describe all organisms as self-assembling, self-maintaining, self-reproducing information-processing machines.[13] My point in making this observation about the theoretical tendencies of molecular biology is not to suggest that the 'information' theoretical view is a distortion of what is 'really' going on in nature (about which I do not in any case have any privileged information), nor to emphasise particularly that this schema must be an inexact and provisional representation, which is a point of view that most scientists would accept as philosophically correct but practically irrelevant. My point rather is to indicate the centrality of the concepts of 'information' and 'program' within molecular biology and to draw attention to the fact that our current rules of biological thought direct us to a specific level or mode of interaction in living nature. I want then to focus on this as a historical issue. What molecular biologists have created is a very general approach to the formulation and analysis of biological problems. In some ways we could describe it as a sort of metabiology, as a reductive, information-theoretical idiom in which to think about biology. I want to think about the history and significance of this conceptual move. How has it come about that an avant-garde in biology has created a new language for analysing nature? What forces lie behind the apprehension of nature as programmed matter?

Similar questions can also be raised about our changing conceptions of life. A significant theme in the past history of the life sciences has been the problem of how to classify the world into those parts that are living, or manifest the phenomena of life, and those that do not. Since the eighteenth century this problem has been transformed, through the development of an evolutionary perspective, so that the nature of life becomes an issue in its own right to be confronted scientifically. Foucault presents this transformation, itself part of a very general cognitive reordering, as part of the move from natural history to biology as a life science.

> To establish the great, unflawed table of the species, genera, and classes, natural history had to employ, criticize and finally reconstitute at new expense a language whose condition of possibility resided precisely in that continuum [of species] . . . Up to the end of the eighteenth century in fact, life does not exist: only living beings. These beings form one class, or rather several classes, in the series of all the things in the world; and if it is possible to speak of life it is only as of one character . . . in the universal distribution of beings.[14]

Foucault goes on:

> In the late eighteenth century, a new configuration was to appear that would definitively blur the old space of natural history for modern eyes, . . . we see criticism displacing itself and detaching

> itself from the ground where it had first arisen . . . and during the same period life assumes its autonomy in relation to the concepts of classification . . . and becomes one object of knowledge among others.[15]

Life, released from the confines of taxonomy, becomes a focus for and object of Reason. In our own century this deliberation has continued, within the framework of a biology ordered by the principles of evolution by natural selection and the laws of genetics. For example the Russian biologist Alexander Oparin has suggested that life is the historical condition of matter under which evolution by natural selection occurs.[16] This powerful idea, conceived holistically, that is to say, rendered in terms that stress the properties of wholes and systems, underlies his important work on the origin of life.[17]

At the same time another scientist also inspired by Marxism, the American geneticist Herman Muller, pursued the theme in a reductionist way in a famous article on gene structure written in 1924.

> We are therefore practically driven to the conclusion that the extraordinary property of protoplasmic growth, or reproduction . . . did not exist before the property distinctive of the gene existed . . . The gene, then, arose coincidentally with growth and "life" itself.[18]

He went on:

> . . . In this process those mutant genes whose byproducts (end-products other than their own material not originally necessary for "life") were most useful in further reproduction differentially survived, multiplied and mutated again. On this view then. . . . the great bulk, at least, of the protoplasm was, after all, only a byproduct, originally, of the action of the gene material; its "function" (its survival-value) lies only in fostering the genes, and *the primary secrets common to all life lie further back, in the gene material itself*. [My emphasis, EY][19]

Muller thus reduces the question of 'life' to the problem of gene structure; those properties, patterns and relations which are characteristic of the evolutionary process of life coalesce in or are derivable from the specific structure of the gene. Carlson and Roll-Hansen have argued very persuasively that this idea was of fundamental importance to the early deliberations in molecular genetics.[20] For Muller, then, life inheres in the gene and the gene concretises life.

Forty years later the Marxist crystallographer J.D. Bernal

returned to the same theme in his book, *The Origin of Life*,in which he wrote:

> The work of the biochemists and later of the molecular biologists has shown how the appearances of life are biological and scientific consequences of the determination of intimate structures down to molecular and atomic level . . . the new knowledge of how the molecular mechanisms of the main elements of reproduction took form in the first place is completing the work of molecular biologists in realising how the functions have been carried out in natural systems throughout the whole of cosmic history—life is beginning to cease to be a mystery and becoming practically a cryptogram, a puzzle, a code that can be broken, a working model that sooner or later can be made.[21]

Bernal goes on to sketch out a theory of the molecular immanence of life, which he defines as 'the partial, continuous, progressive, multiform and conditionally inter-active,self-realisation of the potentialities of atomic electron states'.[22] This ambitious, partisan condensation of life into an elaborate definition must have startled many classical biologists, but as an exercise in strategic abstraction it is a familiar move in molecular biology. One finds similar observations in the history of the concept of heredity written by the French molecular biologist François Jacob, *The Logic of Living Systems*, where life as currently conceived is presented as a logic immanent in living matter, an order disclosed by and for an age of computers, complex electronic systems and networks of communications.[23]

I believe that the interesting feature of these general conceptions is their historical relation to structural change in biology, to its institutional reordering and gradual translation into more subordinated forms. This is not to suggest either that these ideas about life are simplistic, or misleading, or incorrect or that they are *merely* devices or weapons to facilitate intellectual change. Clearly they are valid representations of nature *and*, I shall suggest, the means through which the reorganisation of biology has been thought, argued, promoted and resisted. Concept, strategy and institution have been in interaction in a process of change which I explore in the pages that follow.

Incorporation

The picture of molecular biology that I want eventually to create places it as a set of concepts and an approach in an important, organising position within a complex of biomedical research institutions. Its history, I shall suggest, should be seen as the working out of indirect, highly mediated processes of direction, through the constitution of new disciplines and areas of research, and the re-arrangement, reconstitution and re-alignment of old ones.[24] This process of incorporation involved building specific assumptions, implied goals and concealed relations of control into new sciences so as to facilitate the progressive assimilation or annexation of biology by capital, a situation that is now clearly manifest in the new forms of biotechnology.

It is important to emphasise that this is not a process that has just been triggered from a state of rest, or set in motion by a particular economic crisis. But in suggesting that the incorporation of molecular biology is coextensive with the history of the discipline, I do not want to overplay the strategic foresight of any of the historical actors. My account tries to display the continual process of setting limits and guides, redefinitions of the scope, style and purpose of biological inquiry, reorientations of scientific outlook, the breaking up of old forms of work and forging of new modes of collaboration, training and advancement, and the protection of new styles of research, within a changing system of highly mediated control by capital, exercised by private, corporation-based foundations, the State and industry.

As Foucault puts it in talking of the relation between 'regimes of truth' and power:

> The problem is simultaneously to distinguish different events, differentiate the networks and the levels to which they belong, reconstruct the links between them and see in what way some grow out of the others. It is at this point that I reject those analyses which concern the symbolic or the area of significant structures and instead turn to analyses that are undertaken in reference to the genealogy of relations of force, strategic developments, and tactics.[25]

Three further observations flow from the reference to force and strategy. Firstly programmatic change in biology has been thought through and argued for in terms that make no explicit reference to structures of control. The process has been one in which a certain alignment of interest has come about between an emergent disciplinary elite and various fractions of corporate capital, and I will attempt to construct this hidden dimension from the strategic moves of the various parties involved. Secondly the indirectness of the process of incorporation is such that scientists believe they retain — and do retain — a degree of autonomy over the planning of research. The historical evidence suggests unambiguously that this autonomy is being continually eroded and that it has always been exercised within an institutional and cultural space defined by political and economic power, so that I suggest that we think of research as being carried out in a condition of 'directed autonomy'. Thirdly incorporation in relation to scientific and technological research is usually taken to refer to institutional, technical and cognitive reorientation, so that the fruits of science become more immediately applicable to capitalist production, and less open to contestation.[26] The implication is of a gradual shift from basic to applied research, a change in the type of institutions where such work is carried on and in their financial support and degree of managerial control, and a revaluation of the criteria used to select areas for investigation. But incorporation is also a term used by various cultural critics to indicate the appropriation, transformation and neutralisation of radical and revolutionary initiatives in the arts or in cultural forms other than science by the hegemonic institutions of bourgeois society, such as the press, the theatre, TV and the film industry.[27]

Something similar has happened with molecular biology. A significant number of the scientists who assisted in the creation of the conceptual and methodological ground-rules of the subject were political radicals — J.D. Bernal, Herman Muller, Linus Pauling, Salvador Luria, Jacques Monod amongst others — whose commitment to the re-casting of scientific theory gained strength and purpose from their politics and the philosophical systems to which they adhered. Nonetheless what was once a challenging, even revolutionary, advance in biological thought is now an important form of support to capitalist medicine and agriculture — its subversive implications tamed and its profitable applications managed under corporate control. Thus if we are really to understand incorporation we have to consider not only problems and goals, but also concepts, methodological assumptions and the control of the meanings which become attached to scientific ideas, and to specify how scientists' activities and plans evade, dovetail with or catalyse this historical process of incorporation.[28] Simon Pickvance's sensitive account of his work as a molecular biologist is an interesting illustration of this process and the struggle to resist it.[29] He has described the fascination of microscopic study of embryological development, especially the symbolic power and emotional undertones of the process he was observing, which came into conflict with the technical dictates of his research, the economics of grant support and the social consequences of the organisation of work in his laboratory. Work as a research student in a prestigious, highly competitive molecular biology institute required him to banish certain pleasures from his work, to efface aspects of himself, to cope privately with the strain of these acts of repression, and to accept the hegemonic forms of a reductive, managed biology in which access to further opportunity is exchanged for results. In the end he opted out of this struggle and retrained as a craftsman in the building trade. One of the many interesting aspects of the article that resulted is the indication that his alienation has a historical character: it belongs to a specific era in the development of biology as labour.

Ideologies of Practice

Pickvance's article is one illustration of how people react to the process of socialisation as a research scientist at particular, given moments in the history of science. Manifestly the practice of science changes continually, and scientists are correspondingly required to re-think the meaning of what they do, in theoretical, technical and institutional terms, as theories, skills and places of work are reordered and reconstituted. One can say that all scientists learn to understand the meaning of their activity by thinking in the terms of a structured, encoded representation of that activity. Luke Hodgkin has suggested that such representations be called 'ideologies of practice'.

> I'd rather say that scientific work, like any other work, takes place within the framework of an "ideology of practice" — an image of what that work is, what its significance is, which questions are interesting and which approaches to them will give results; in short a complex of representations of scientific activity. And this ideology — shared to a greater or lesser extent among a number of scientists — is socially determined, and in its turn determines the choices and orientations for a scientist's practice, and so for the *kind* of knowledge he produces . . . A scientific ideology of practice, is, in today's conditions at least, a bourgeois ideology. By this, I mean that one of its functions is to depict the activity of doing-science-within-capitalism as an acceptable choice for the individual.[30]

The suggestion is that such ideologies are not simply depictions of the practices of particular individuals, disciplinary groups or professions. They are encoded versions of the form, scope and purposes of specific practices which purport to explain the significance of those forms of activity, to 'make sense' of them by fitting them into a comprehensive system of meanings. Thus Stuart Hall, in writing of the relations between culture, consciousness and ideologies states:

> . . . There is also the distinct area in which classes "experience" their own practice, make a certain kind of sense of it, give accounts of it and use ideas to bring it into a certain imaginary coherence . . . To "give sense" in this way is, fundamentally, to locate oneself and one's experience in the already objectivated ideological *discourses*, the sets of ready-made and preconstituted "experiencings" displayed and arranged through language which fill out the ideological sphere.[31]

He goes on:

> . . . ideology as *a social practice* consists of the "subject" positioning himself in the specific complex, the objectivated field of discourses and codes which are available to him in language and culture at a particular historical conjuncture — what C.Wright Mills calls "situated actions" and "vocabularies of motives".[32]

Clearly scientists individually and collectively draw upon such vocabularies. Mulkay for example has argued that leading scientists deploy particular images and descriptions of science in an attempt to defend their autonomy as a managerial elite.[33] This seems to me an accurate perception and one which can be expanded to include the sense indicated by Hall. That is, that scientists, in thinking in the terms of the dominant professional ideology, 'misrecognise' their situation, or accept 'false inflections' — the terms are Hall's — of the social relations of scientific research. Thus they underestimate and deny the ways in which their autonomy is directed, ordered and corralled.

But ideologies of practice also have another dimension, as Hodgkin makes clear in linking together the practice and object of research. How you study something is in part determined by what you believe the structure, boundaries and organising relations of that object to be. I want therefore to extend the conception of ideologies of practice to include cognitive principles and epistemological conventions. In a sense this is implied by my earlier discussions of conceptions of life and nature, and their role in the organisation of biology. For the struggle over how to do biology, which constitutes its history, has been a conflict between competing strategies for research, over alternative institutional moves, over what techniques and technologies should be imported, so as to produce useful and reliable knowledge — and at the same time over how life, organisms, heredity and evolution should be conceived so as to facilitate the production of knowlege in specific forms.

THE REDUCTIVE SHIFT TO MOLECULAR PROCESSES

Not only can one construct a history for molecular biology as a series of discoveries, experiments, models, concepts and generalisations, a path through the double helix, to transmute the title of a recent historical monograph on the subject,[34] it is also possible, and important in my view, to describe its place in the more general historical project of modern science, which I take to be the unevenly progressing, frequently contested, tactically varying reduction of processes and relations in nature to matter, motion and number. The significance of molecular biology derives then from the generality of the phenomena which it seeks to apprehend and from its consonance with and consolidation of a historical current of explanation. Molecular biology registers, and has served to promote, a shift in what counts as knowledge, in what features of living matter biology should seek to abstract and explain and what is acceptable or interesting as an explanation. It represents a movement beyond the earlier 'intermediate' disciplines such as physiology — or (as Bob Young has put it) way-stations of the reductionist programme — to the reformulation of biological properties and functions in terms of molecular structure.[35]

One can view molecular biology as one of the latest stages in an extended process of framing the manifold of nature in specific ways, so as to constitute problems for study, to define the terms and conventions for meaningful and useful scientific analysis and to facilitate the jettisoning or suppression of what came to be seen as archaic forms of thought. This movement, with its roots in the sixteenth and seventeenth centuries, began with the elevation of a mechanistic conception of nature from which all natural tendencies, immanent causes, vital principles and affinities have been removed and which focussed specifically on those attributes which can be reduced to the properties of matter in motion. It progressed through the idea of a biological property in eighteenth century physiology, which can be seen as a compromise between the ambitions of reductionism and intractability of living matter, and continued with the systematic classification of living things which burgeoned in the nineteenth century into a theory of the relatedness and development of living nature, based on the ideas of adaptation, competition and differential rates of reproduction and survival. This expanding programme of reductionism accompanied the successive disclosure of different orders in nature. The latter idea is presented very clearly in François Jacob's history of heredity, *The Logic of Living Systems*, where he writes:

> There is not one single organization of the living, but a series of organizations fitted into one another like nests of boxes or Russian dolls . . . Each level of organization thus brought to light leads to a new way of considering the formation of living beings. After the sixteenth century, a new organisation is seen to have appeared on four separate occasions, each time a structure of a higher order: first, at the beginning of the seventeenth century, the arrangement of visible surfaces, which might be called the first-order structure; then at the end of the eighteenth century, the "organisation", the second-order structure which underlies organs and functions and was finally broken down into cells; next at the beginning of the twentieth century, chromosomes and genes, the third-order structure hidden in the heart of the cell; finally in the middle of the present century, the nucleic acid molecule, the fourth-order structure, which is considered today to be the basis of each organism, of its properties and its permanence throughout generations . . . the successive disclosure of these structures . . . is the way in which "generation", a new creation each time and always requiring the intervention of some external force, has become "reproduction", the intrinsic property of all living systems.[36]

Within this organised set of theoretical structures, Jacob then discerns enduring themes or organising principles, such as the theory of evolution, which he considers to be the most important since it now unites all the disciplines concerned with living beings and provides a causal explanation of the living world and its heterogeneity. Evolution can now be theorised as a complex, densely structured unity, of which many separate aspects can be isolated for specialised study, but always in terms which depend upon and support the accepted explanation of the historical development of living matter, namely the neo-Darwinian theory of evolution by natural selection. Jacob also emphasises that the theory of evolution has incorporated (or has assimilated and given additional meaning to) a number of unifying principles, such as the cellular composition of all organisms, or the universality of the genetic code or the essential similarity of the biochemical reactions which mobilise energy in living systems.[37] But within this increasingly elaborate and powerful explanatory schema, different features or processes take on a central importance. Thus the twentieth century has seen increasing attention given to the lawfulness of the process of heredity, with the emergence of the science of genetics. In the 1930s the success of genetical explanations, by then extended to the level of populations, was a major source of support for the reconstitution of evolutionary theory as neo-Darwinism. The accounts by Jacob and others suggest then that one should regard the evolutionary process as the background of twentieth century biology, a level of reality to which all analysis must at some point be referred back. Meanwhile, more analytic and reductive studies, such as work on the structure of the gene, have been placed in the foreground, and have been cast in the role of more important and more illuminating studies, as the leading edge of reductionism.[38]

One rationale for taking this view appears in the passage by Herman Muller quoted above. For him the focus of the most advanced form of biology should be on the structure of the gene, since, he maintains, the operative principles of evolution and heredity are located there. Muller goes beyond the abstract formulations of classical genetics to regard the gene as a molecule, as a substructure of the chromosome, and places it in a conceptually central position in biology. Others, like Muller's erstwhile senior colleague, Thomas Morgan, took a more cautious view, although he eventually consented to seeing gene structure and the transmission of hereditary traits between generations as problems to be tackled in isolation from developmental issues. The progress of biology in general was less rapid because it was committed to explaining the complexities of development.[39] Still others — like Linus Pauling, William Astbury, and J.D. Bernal — were fired by a vision of what a more general knowledge of molecular structures could achieve for the understanding of biological function, although within this programme the structure of the gene clearly had a special place.[40] But despite their differences, these positions from the pre-history of molecular biology share a common derivation from a reductionist tradition, by virtue of their commitment to levels seen as more fundamental than those attained by classical genetics or biochemistry.

Three features stand out in the development of this movement in the 1930s — firstly the de-emphasis of natural historical, evolutionary or functional explanations by a nascent group of biologists,[41] secondly the stress on simplicity in nature and simplification as a theoretical gambit,[42] and thirdly the examination of process via the analysis of specific substances and their structure. An example of this last tendency is to be found in the work of Oswald Avery at the Rockefeller Institute in New York, an institution firmly committed to a mechanistic experimental biology.[43] Avery investigated the phenomenon of transformation in bacteria, which appeared analogous to and was then shown to be identical with genetic change. In a long series of experiments he attacked this problem by concentrating on the isolation of the substance responsible, which in 1944 he showed to be deoxyribonucleic acid (DNA). One implication of this tremendously important result was that genes were made of DNA, not of protein as hitherto believed; another was that it made sense to do recombination experiments with bacteria, to gain more knowledge of the structure and organisation of their hereditary material. Max Delbrück, Salvador Luria and Alfred Hershey and the younger scientists they began to train all took up this opportunity very quickly, exploring the implications for molecular genetics, whilst the biochemist Erwin Chargaff and others began refining the latest analytical techniques to learn more of the chemical structure of DNA. At the same time, in the late 1940s, work on the molecular structure of protein and DNA was also being pursued, so that in 1952 and 1953 Watson and Crick were able to assemble enough evidence to support the daring conjecture that DNA had a double-helical molecular structure and that its involvement with the process of heredity should be derived from its structure. This idea, received enthusiastically by a small network of scientists, and rather more sceptically or uncomprehendingly by more classical biologists, opened up further lines of investigation, including work on the genetic code and the regulated production of proteins from encoded hereditary information. By the late 1950s the analysts of protein structure were beginning to have some success and there emerged a number of points of closer contact between the informational schools in molecular biology. In the early 1960s Jacob and Monod proposed a very powerful model for the genetic control of protein synthesis in bacterial cells, and by the mid 1960s the main outlines of the genetic code had been cracked after furious competition. With these successes behind them, molecular biologists began to claim in a more confident and militant tone a general relevance for their concepts and analytic approach throughout the life sciences. Thus for example in 1970 Jacob wrote of the concept of the genetic program as uniting hitherto divergent traditions of explanation, of a theoretical reconstitution that has produced, and can take credit for, a new vision of unity.

> The two kinds of order meet at the level of heredity, . . . so to speak. If the species are stable, it is because the programme is scrupulously recopied, sign by sign, from one generation to another. If they vary, it is because from time to time the programme changes. On the one hand, it is necessary to analyse the structure of the programme, its logic and its execution; on the other, to examine the history of programmes, their drift and the laws governing their changes throughout generations in terms of ecological systems.[44]

The tone here is carefully expository and permits various inferences about the importance of molecular biology. In other contexts molecular biologists have sought to exploit more aggressively the implications of their theories and methods. Thus some molecular biologists, Francis Crick being one distinguished example and Jacques Monod another, have developed a special deductive, simplifying, theoretically-inclined style of reasoning. That is made apparent in Robert Olby's memoir on Crick,[45] and in Horace Judson's history of molecular biology, where Crick is shown vigorously exhorting rather sceptical virologists to rethink their problems in terms of molecular structure and crossing swords with biochemists over how to analyse the problem of the genetic code.[46] James Watson is renowned for much more trenchant dismissal of the methods of other biological disciplines, and for some people this arrogant vanguard mentality has had great appeal.

Molecular biologists aspire, then, to a more strenuous style of reasoning than those used by their colleagues in other disciplines. Sydney Brenner has spoken, for example, of a cult of minimal experimentation.[47] Yet molecular biology is not theoretical biology, even though it is closely integrated with biological theories of general scope. My suggestion is that it offers a general language, or set of rules, in which to think about biology from a genetic standpoint, by reference to the concept of the genetic program and a vocabulary of systems-theoretical ideas, such as 'feedback' and 'message'. James Watson's *The Molecular Biology of the Gene*, a major textbook of its time, may be read in its successive, accreting editions since 1965 as programmatic in two senses: first, as an exposition of the program concept; and second, an indication of how the problems of various areas of biology can be attacked programmatically from the molecular level, in novel reductionist ways. Leading molecular biologists and their recruits have sought to colonise certain areas since the mid-1960s, e.g., embryology, immunology, neurobiology, and tumor virology.[48]

Molecular biology thus represents a bold initiative in the life sciences, with its roots in structural chemistry, biochemistry, X-ray crystallography, genetics and bacteriology in the 1920s and 1930s. Its practitioners were at first a network of researchers in different fields, some of them in marginal and precarious positions, and latterly a highly regarded, well-connected scientific elite, training and controlling a large group of aspirant researchers. They have gradually learnt to recognise, articulate and propagandise the potential of their subject, and sought the power to reorganise biology in the face of considerable criticism. Arguably they have done this by forcing through changes in what counts as an explanation in biology and how problems and hypotheses should be formulated. But this process should not be seen simply as one of their own making, that is, as the rational exploitation of the levels of order they happened to discover in nature. The history of molecular biology, as I try to suggest in the following sections, is also a social history, a history of changes in the social relations of science, a complex process with both conceptual and institutional moments.

SOCIAL RELATIONS: FOUNDATIONS, STATE, INDUSTRY

The closing decades of the nineteenth century saw major changes in the economic and social structure of the major industrialised and rapidly industrialising nations, such as Britain, France, Germany, the United States and Japan, as economic power became increasingly concentrated in large industrial corporations. Such enterprises required an expanding clerical labour force to service their administration, greater numbers than hitherto of skilled engineers and scientists to develop new products and to reorganise and improve the reliability, scale and profitability of production, and a new class fraction of executives to manage the operations of these complex business institutions. In the political sphere this economic expansion produced changes in the structure of the State in the same period, with a strengthening emphasis on institutions and policies that were intended to assist the longer-term survival and expansion of capitalism, at the national and international levels. This interventionist role, and the coercion, negotiation and management it implies, necessitated a certain autonomy from the economically dominant class so that the process of accumulation could be sustained and the legitimacy of class domination protected. Some Marxist commentators on the State have written of the emergence this century of a 'class conscious political directorate' within the State machine.[49] Others have stressed the importance of challenges to State power in forcing reorganisation and new forms of domination.

> ... it is the threat of a powerful working-class movement which galvanises the ruling class to think more cohesively and strategically and to restructure the State apparatus to this end. Those countries which have experienced strong centralised challenges to the power of the capitalist class are those which have developed a unified State apparatus to counter those challenges ... The responses also differ, but all involve a move towards a more centralised interventionist State.[50]

Whether or not these analyses are accurate it seems to me important to note the continual struggle to restructure capital, labour and the institutions of the State, to protect the circulation and accumulation of capital — a struggle engaged in by competing class fractions, interest groups, leaders of professions and reform-minded civil servants. In the fields of science, medicine and technology the promotional activities of the State have clearly become extremely important in the twentieth century. However, in reviewing the institutional history of science what stands out is the time taken for a strategic sense to develop, gain acceptance and lead to structural change, the reluctance of leading scientists to conceive their practice as a long-term mode of support to industrialisation and economic expansion, and the different phases through which the direction of science and technology has passed. Significant changes have been triggered by world wars, and the role of private foundations, such as those using the capital controlled by John

D. Rockeller of Standard Oil and Andrew Carnegie of U.S. Steel respectively, has often been to force through innovation against cultural, political or professional resistance.

The latter half of the nineteenth century saw a growing interest in applied industrial research, particularly in the chemical and electrical industries, and the establishment of new industrial research laboratories for product and process development.[51] The period also saw increasing resources being given to the State research institutes to promote economic expansion. Similarly the founding of the Kaiser Wilhelm Gesellschaft in 1912 in Germany was an attempt by state and industry to improve national industrial competitiveness through the support of fundamental research. But the measures taken in different countries to promote these aims varied considerably, with the result that the First World War produced a scientific and technological crisis that led to the creation of new institutions such as the Research Councils in Britain and opened up a space for entrepreneurial scientists, such as George Ellery Hale in the United States, to lobby government for more resources. In America particularly, the decade of the 1920s was a period of academic expansion, with massive endowments going to the universities to open new departments, increase the production of students and extend the time and space available for research. Indeed Coben has suggested that the provision of research fellowship schemes with foundation money in the 1920s was of major importance scientifically in allowing scientists time away from teaching, in increasing their mobility nationally and internationally, and in allowing an intensification of research by an increasingly confident, ambitious profession.[52] But the general level of State support for scientific research at this time was low, and the control of funding in universities rested largely with university presidents and research administrators, rather than with national agencies or nationally organised groups of leading scientists as is now the case.[53]

In addition to the changes in science and technology, the late nineteenth century also saw changes in the nature of medical care and education, through a development of the technical resources that doctors could deploy, for a changing clientele from expanding fields such as bacteriology, biochemistry and immunology. Just as private endowment and philanthropy had a marked effect on higher education in science and technology, so also in medicine foundations like those established by Carnegie and Rockefeller to use their funds to assist in the 'rationalisation' of capitalist society had their counterparts in America and elsewhere through the promotion of scientific medicine, the establishment of research institutes (such as the Rockefeller Institute for Medical Research in New York), the closure of low-status medical schools, and corresponding pressure on doctors to spend more time on research — a move which the profession struggled hard to resist. Just as Standard Oil eliminated 'inefficient competitors' in the oil business, so the Rockefeller philanthopies drove out the competition in scientific medicine. The less professional and smaller units were driven to the wall. Robert Kohler notes that, according to the Rockefeller Foundation's own documents,

> It was the policy of the Foundation to "make the peaks higher", to help strong institutions to serve as regional models, and to allow normal competitive pressures to reform the national system by eliminating the weak and stimulating the more progressive.[54]

Thus by the early 1930s in America, the Rockefeller Foundation trustees and officials had gained some years of experience of supplying money for university development and acting to 'rationalise' health care — levelling out objections, negotiating with professional groups, setting up new institutes and hospitals, imposing new standards of recruitment on the medical profession, installing systems of management, organising research and public health programmes on an international basis and promoting new scientific specialities. Moreover such strategic initiatives were more innovative and more generously supported than those undertaken by the state in America, given the ways in which governmental responsibilities for higher and professional education and research were construed at this time.

> The ratio of private-to-federal expenditures for medical research continued relatively unchanged for the first four decades of the century. And it was accepted as natural that the private side should dominate . . . It could even be said that the private sector was expected to support medicine and medical research.[55]

Similarly in Europe no governmental agency had the political will to play such a powerful role in biomedical research, and no private body the necessary economic power. Kohler has discussed, however, the structural changes implied by the formation of the Medical Research Council in Britain in 1920, with a division of labour agreed with the Ministry of Health which allowed a national research policy to be carried out under the control of a medical elite.[56] Against this background of the 'rationalisation' of private health care and intensification of biomedical research, the Rockefeller Foundation decided in the early 1930s to pursue a new strategy in research support, which in due course effectively constituted a new discipline in the life sciences.[57] In 1938 this programme was named 'molecular biology'; this initiative was of decisive importance and is analysed below. But it is also important to realise that these moves by the Rockefeller Foundation emerged from a period of reorganisation to produce a coherent, more effectively managed set of institutions. The proposed elitist support of highly specialised research — the policy of 'making the peaks higher' — was not accepted by some of the Trustees or by some leading scientists. Indeed one theme that emerges clearly from Robert Kohler's impressive analysis of the foundation policies and debates in the 1930s is the tactical skill with which various proponents of strategic, more directive funding of more specialised research developed support for their policies against the opposition of those who adhered to more traditional ideologies of practice that allowed researchers and universities more autonomy to plan and work as they wished.

Nonetheless in the universities some academics were seeking to build up institutions which embodied new conceptions of research. Through the 1920s George Ellery Hale, Arthur A. Noyes and Robert Millikan worked to develop the California Institute of Technology in Pasadena. They created a small, well-endowed university with a deliberate concentration on research. Its commitment to the most advanced investigations attracted leading scientists. In 1928 they recruited the geneticist Thomas Hunt Morgan from Columbia to establish a new Division of Biology.[58] In the early 1920s Morgan had already become a major figure in the establishment of the chromosome theory through the experimental breeding of *Drosophila* fruit flies. In the early 1920s he and others had striven to develop a new style of research in the life sciences — empiricist, experimental, anti-speculative, vigorous in method, analytic rather than taxonomic and mechanistic in orientation. This was to be a new 'American' biology, in contrast with the perceived vagueness of much European work of the late nineteenth century.[59] In Morgan's view the chromosome theory was built up on just this basis. With this success behind him, he moved on to Caltech with the intention of creating a new type of biological institution — research-oriented, concerned with general, fundamental studies, based on the exact methods of physics and chemistry, staffed by the best recruits available for a multi-faceted programme of research, designed to encourage collaboration and interaction as much as possible, and with a coherence derived from a specific conception of life. Morgan described it thus:

> It is with a desire to lay emphasis on the fundamental principles underlying the life processes in animals and plants that an effort will be made to bring together, in a single group, men whose common interests are in the discovery of the unity of the phenomena of living organisms rather than in the investigation of their manifold diversity.[60]

In the years that followed, the Division of Biology built up a reputation for experimentally-based, collaborative, reductive research that drew in any visitors and attracted generous financial support, particularly for arrangements which allowed physical and life scientists to align their research programmes. Much of that money was channeled to the geneticist, George Beadle, by foundation officials like Warren Weaver, head of the Division of Natural Sciences of the Rockefeller Foundation from 1932. It is Weaver's innovative activities that I now want to consider by drawing on, reinterpreting and adding to the work of Robert Kohler, who has directed our attention to the role prepared for Weaver by the development of the social relations of science.[61]

It is important to bear in mind that Weaver's moves were made against an initial background of economic recession. At a time of massive unemployment, which included qualified scientists and engineers, the closure of industrial plants and collapse of businesses, attacks on academic research from Fascist governments and parties, and major political struggles over economic strategy, the Rockefeller Foundation was able to appropriate $10 million per year, of which about $7 million was divided between the social sciences, medical, humanities and natural sciences divisions. What Weaver did was to develop a system of patronage and direction of research, i.e., to bring modern management to science. Many of the more distinguished scientists with whom he dealt were also engaged in lobbying for greater governmental interest in science, for example through the Science Advisory Board of the Roosevelt Administration.[62] There was already a growing strategic sense on the part of industry and government in the 1930s, which was illustrated by the increase in the number and size of industrial research laboratories[63] and by the plans laid by Public Health Service officials for biomedical research. There was also growing public interest in targeted medical research, represented by the foundation of the National Cancer Institute in 1937[64] and the funds gathered to fight poliomyelitis. However, it took the shock of the Second World War to realise the institutional and economic possibilities envisaged in the 1930s by the lobbyists for science in the US, Britain, France and elsewhere.

Managed Science: Warren Weaver

Warren Weaver was born in 1894, the son of a druggist of German extraction in a small Wisconsin town.[65] He emerges from his memoirs as a thoughtful, religious, meticulous person, deeply committed to science intellectually, spiritually, and vocationally. He became a well-liked, trusted figure in the higher circles of American science. He was a powerful and efficient administrator, well able to respond to political and institutional cues and shifts of emphasis, even if, according to his own estimation, he was somewhat unimaginative. He began his university career as a civil engineer but soon changed to mathematics and mathematical physics, subjects which he then began to teach. He taught first at Throop College, soon to be renamed the California Institute of Technology, and then at the University of Wisconsin at Madison, where he began to collaborate with the physicist Max Mason before Mason moved on to the Presidency of the Rockefeller Foundation.

By 1928 the growth of American universities relative to the endowment resources of the Rockefeller institutions, together with their own disordered growth, forced a reorganisation directed by Raymond D. Fosdick — J.D. Rockefeller's chief counsel, a League of Nations advocate, and a Progressive reformer. The result was a single Foundation with five divisions and a new strategy of concentrating effort on science and supporting whole areas of learning, unified by their relevance to a programmatic 'science of man'. However, as Kohler points out, concentration entailed making choices, setting priorities among disciplines and even within disciplines on the basis of Foundation priorities. Officers became, in effect, policy-makers for scientific research. Into this role Weaver stepped in 1932, having been invited by Mason to assume the Directorship of the Natural Sciences Division at a time of re-appraisal and a move away from funding the physical sciences, which had been generously supported in the 1920s. Weaver, an outsider to biology, put forward a proposal for the programmatic support of work on 'vital processes' of long-term significance to medicine which was to be part of a more general, Rockefeller-assisted 'science of man'. Weaver also intended to catalyse certain changes in the life sciences, to emphasise the value of exact, analytic, vigorously formulated, reductive experimentation, based on the methods of physics and chemistry.

> . . . the phrase "vital process" . . . was the label which I wanted to have tied to the program because I wanted to get it away from any ancient limitations adhering to the word "biology" . . . And I wanted to say that this was a science of vital process; this was a procedure in which you bring to bear the whole battery of science to the study and investigation of the phenomena of living things. But for some reason or other, some of my biological friends found this phrase "vital process" a little precious. The trustees never quite understood whether they were being sold a bill of goods or not. [But] it *was* biology, in its purpose; it was biology in its ultimate orientation. It was not by any manner of means biology in terms of classical technique.[66]

Gradually Weaver found his feet in his new, prestigious, well-paid job and began to cope with the struggles over policy. In early 1933 he set off on a European tour of reconnaissance, the first of many, the results of which were entered up in a diary that could be circulated within the Foundation. In later years Weaver stressed just how much time was spent travelling, and how much care was taken to solicit leading scientists' views by strictly managed personal contact in discussions of their own work, the prospects for their field as a whole and the potential of younger researchers. In the 1940s Weaver wrote a memorandum or a handbook of etiquette for junior Foundation personnel on the management of the complex relations of power, patronage, acquaintance and indebtedness. The object, it is clear, is to retain control in the hands of the Foundation, so that its own goals can be pursued, and its own principles and priorities implemented, whilst channelling support to those deemed worthy of long-term support.

> . . . In the private foundation, the responsibility, I think, is to support excellence; it is to support those who are already good and make them better, so that they may be scouts way out ahead, really penetrating the frontiers of knowledge, really opening up new opportunities for the future.[67]

In a critical discussion of the later bureaucratic practices of the National Institutes of Health (the chief post-War governmental sponsor of American biomedical research), he also said that

> . . . the only way to judge a proposal is by virtue of the intellectual capacity, the record of imagination, of dedication, and of interest of the man who proposes to do this work. The specification of the problem is almost always unimportant . . . [And] the way to find out whether a man deserves support [is] to go and sit down on a stool in his laboratory with him . . .[68]

The emphasis is on the *right* kind of controlled personal contact, leading to an interlocking system of evaluations, to enable a programme of research support to be built through a series of informed choices between alternatives. Weaver indeed stresses the necessity for large foundations to operate flexibly within the clear outlines of a programme.[69] He makes clear that his own constant tending of the programme distinguished him from foundation officials like Fred Keppel of the Carnegie Corporation who operated in a much more informal, unsystematic, eclectic way.[70] Weaver, by contrast, was tremendously conscientious in developing skills to perform the new task of managing science, without appearing to do so.

After a tentative start, with support from the programme of psychobiology run by the Medical Sciences Division, Weaver's plans were endorsed in 1934 by a Committee of Appraisal composed of distinguished scientists with some criticism of psychobiology. His confidence in his new role as manager increased. A new fellowship scheme in 'vital processes' was set up, and Weaver became more active in the selection of grants; he made changes in their administration by placing emphasis on planned or programmatic projects, usually involving application of some physical technique to biological problems and often involving collaboration of biologists and physical scientists.[71] By the late 1930s, group grants, of which the Division of Biology of Caltech was a favoured recipient, also began to play an important role in Weaver's programme. He saw them as pilot projects for future capital grants to university departments for the purpose of institutionalising the application of physical science to biology on a permanent basis. Clearly then Weaver was consciously playing a strategic role. He was facilitating technical and methodological change; making resources available for selected projects, programmes and institutions; imposing a preference for biological work informed by physics and chemistry, a choice that drew support from the example of the reductionist programmes at the Rockefeller Institute for Medical Research in New York[72]; emphasising new ways of more specialised, collaborative working; and assisting the development of new research technology into a programme that came increasingly to emphasise reductionist studies in biochemistry, structural chemistry and genetics. Into this framework were fitted grants to improve the electron microscope and ultracentifuge in order to increase the production and use of radioisotopes in biochemistry and the use of X-ray crytallography to analyse organic materials. Grants were also given to make available these technologies to those forward-looking biology departments that wished to install them as a general resource. In 1938 Weaver re-named his programme 'molecular biology'.[73]

Given the financial power that the Rockefeller Foundation could exert, any programme it supported in the 1930s was certain to have a significant effect. Through his programme Weaver channelled extremely valuable support in various forms to a number of scientists whose work coalesced in the post-war discipline of molecular biology. But the significance of the programme lies not only in its having accelerated a selected number of linked lines of technical and theoretical activity. Weaver's refinement of bureaucratic techniques and his development and application of a new mode of control over science illustrated, as did a number of other programmes, that research could be selectively supported and efficiently managed. When the Second World War began, the leaders of American science presented to the American administration an institutional system with just this capability: the Office of Scientific Research and Development (OSRD), incorporating a Committee of Medical Research, administered the work on the atomic bomb, the proximity fuse, radar, computer-controlled anti-aircraft guns, antimalarial drugs and operations research, amongst other projects. Weaver's expertise brought him the task of running the Applied Mathematics Panel of the OSRD. The War-time programmes of the OSRD had a major impact on the conduct of the war and on subsequent technological, industrial and military development. But also their success — and analogous experiences in Britain — forced major political changes in State support of research. As the War closed, debate began over the scale and nature of future funding and the means of its administration. Its outcome placed much greater financial resources and managerial power in the hands of the elite, which Weaver and others had serviced so efficiently through the 1930s, and the political astuteness of whose members had allowed a consolidation of influence as the War progressed. As Pursell has written:

> . . . the OSRD was not only a success within the narrow research limits it set for itself, it was equally successful in perpetuating the scientific leadership which, coming out of the Depression of the 1930s, built the scientific establishment of the Cold War period.[74]

Post-War Managerial State

Strickland has suggested that it was only with the coming of the Second World War that the US government had a clear, direct, purposeful medical research policy.[75] Between 1941 and 1947 the Committee on Medical Research (CMR) of the OSRD spent $25 million, of which $5-6 million went to research on malaria.[76] In 1944, as Congressional, governmental and public debate began on future provisions for research, the statutes of the Public Health Service (of which the National Institutes of Health had been the major research arm since 1930) allowed it to play the promotional role taken on by the CMR at the beginning of the War. This move was assisted by the revelation of the poor health of many American conscripts in the War. In 1946 leading NIH officials like Rolla E. Dyer and Cornelius Van Slyke began to take over the administration of continuing War-time research through a Division of Research Grants and to develop a rapidly expanding system of grant support, controlled by study sections of leading scientists. Strickland presents this as a pre-emptive response to a situation where considerable growth was clearly going to occur.[77] There was indeed an increase in funding, for PHS research from $180,000 in fiscal 1945 to $4,000,000 in fiscal 1947, and an appropriation for the NIH of $8,000,000 in 1948, which was ten times the 1939 figure. However, the exercise was conducted in a fairly informal way. C.V. Kidd, an NIH official who later developed various managerial and organisational devices to tighten central control over funding, has said:

> As a matter of fact, during the years immediately after the war, on the one hand, the NIH ran two separate operations almost — one was a ruthless propaganda campaign, highly political, marshalling every force to increase the money available. On the other hand, there was an effort to spend the money with a decent, rigorous, scientific examination of what the projects were and who was getting the money . . . There were no management activities. The Division of Research Grants was run really out of Van's vest pocket.[78]

As the budget for the NIH continued to climb, this informality, also to be found at the National Science Foundation, where Paul Weiss set up a 'molecular biology' programme, gave way to a more explicitly managerial approach. This was developed and wielded by a much more aggressive generation of research directors, including people like James A. Shannon, Director of the National Institutes of Health from 1955 to 1968, during which period its budget rose to $1 billion.[79]

Shannon began work in medical research before the War, which he spent working on malaria. This experience intensified his vision of what might be possible with much greater levels of support for research. 'I came out of the war with a keen appreciation of the absolute need, if I was to do what I wanted, of a well controlled clinical service with a very broad support program in basic science'.[80] From industrial work as Director of the Squibb Institute of Medical Research, he moved in 1949 to the newly formed National Heart Institute, one of the parts of the NIH, then to the NIH programme of extra-mural grants, and became Director of NIH in 1955. ' . . . my first responsibility here was to convince an extraordinary Secretary, Mr Folsom, that medical sciences were getting grossly inadequate support.'[81] Marion Folsom, formerly an executive with Eastman-Kodak, a company with a long tradition of intensive research, took office as Secretary of Health, Education and Welfare, with responsibility for NIH. His appointment followed some years of conflict between a Republican Administration and Democratic Congress over budgetary appropriations for medical research, which the latter consistently increased from the estimate set by the former. Shannon wrote:

> . . . I was fully able to appreciate Folsom's attitude that the

> appraisal of science in industry is a sharp one, sharper than it could ever be from the type of support program we have here [at NIH], so I understood his attitude. He said he would like to send for a conference . . . within industry . . . these people who came were hard-headed heads of research and development operations drawn from industry.[82]

They gave their approval to the NIH operation and the budget for 1956 was increased by 30%.

The post-War growth of the NIH as a biomedical research agency with immense resources at its disposal, some of which went to molecular biology, was a complex phenomenon. It has been carefully but incompletely analysed principally by Strickland, Reagan and Rettig.[83] What stands out from their accounts is firstly the role of a group of influential Senators and Congressmen controlling annual appropriations, as well as the activities of very wealthy individuals (like Mary Lasker) in organising lobbies, contacting, retaining and marshalling scientific consultants, carefully dispensing financial patronage to politicians, and building up a system of influence over health and research policy that included an alliance with powerful administrators like Shannon. Secondly they display the attitude of the American Medical Association towards the State provision of medical services, medical insurance and medical education on the one hand, of which they were fiercely critical, and to medical research, on the other, of which they were accepting, perhaps even welcoming.[84]

The effect of the lavish support of research in the 1950s and 1960s was to build up a cadre of highly specialised experts, grouped together in universities and medical schools, and the categorical institutes of NIH, with a broad, long-term relation to hospital-based, technologically intensive clinical medicine, practised and made available within a system which perpetuated massive inequalities of access to health care. The role of the American State, drawing on the promotional lessons of the 1930s and the War years, was to develop medical labour and the practice of research in directions consistent with the balance of political power and its expression in health policy. The State orchestrated this effort through centralised control, a regime of 'directed autonomy', which allowed scientists a space within which to work, whilst policing the ground-rules and boundaries of that activity.

But the growth of the budget also brought growing scrutiny

of how the money was spent and controlled. In the 1960s a series of Congressional Committees produced reports that were critical of the management structure and practices of the NIH; subsequent years have seen strenuous attempts to enhance the mechanisms of centralised control, particularly within the National Cancer Institute. For example one product of continual tension between targeted research on specific problems and general fundamental research has been the institution of 'programming', in the active sense; this entails bringing together a group of scientists who are asked to constitute a programme of research on a topic defined centrally by NIH, but to be pursued scientifically according to a plan laid down by researchers. Hence we have the present structured complex of managed programmes — overseen, funded and directed centrally, through the controlled deployment of scientific advice — a system on which Weaver worked in an earlier phase until his retirement in 1958.

This system has also had important economic consequences for universities and medical schools by increasing their dependence on central funds earned by successful research entrepreneurs, who built up powerful baronies with the money channeled to them not from local sources under local control but by a network of their peers at the national level. Murtaugh, Heirich, Stevens and others have described how research allocations were used covertly to subsidise teaching and building programmes against the opposition of the private medical lobby.[85] Heirich has also described how the availability of money from overheads and non-specific grant allocations opened up for universities the option of investing such money in future research entrepreneurs, whose programmes were to generate further revenues. This system in turn transformed the research practice of aspiring and successful scientists, through an intensification of competition for financial resources, which they were required to manage as an asset in the race with colleagues for results which would secure a further disbursal of funds. Furthermore, access to this system — consolidated by the managerial policies of the granting agencies — was via an increasingly competitive regime of doctoral training and post-doctoral movement around an international network of labs, where the latest skills and techniques were to be picked up. Thus laboratory directors, like James Watson at Harvard and Cold Spring Harbor or Sydney Brenner in Cambridge, have been able to intensify the pace of research and extort greater and greater efforts from young scientists (like Simon Pickvance[86]). Some of these have rebelled, some have suppressed their resentment and hurled themselves into the competition, while some will now take the option of moving into industrial molecular biology.[87]

Restructuring European Research

The account so far has been heavily biassed towards developments in America. This is potentially misleading in three ways: firstly in that the Rockefeller initiative had a significant impact in Europe; secondly in that European researchers have made very significant scientific contributions in molecular biology; thirdly in that national and international plans for biomedical research in Britain are important in their own right and in relation to American activity.

Thus Weaver's programme in molecular biology included support to scientists in Britain like W.T. Astbury, Dorothy Wrinech, Robert Robinson, and Max Perutz and for Svedberg's work on the ultra-centrifuge in Sweden. Weaver made travel funds available for discussions between British, French, German and Danish scientists on the relevance of physics to biology.[88] The Rockefeller Fellowship programme brought European scientists to America, such as Max Delbruck and Jacques Monod who visited Caltech; Weaver in his tours of Europe consulted leading British scientists like F.G. Donnan and A.V. Hill, one of the post-War sponsors of biophysical and crystallographic work in Britain. After the War Rockefeller money allowed French biologists at the Institut Pasteur in Paris to buy equipment and materials, and to hold conferences promoting interest and competence in molecular biology.[89]

In Britain some of the lines of research set up by Weaver were taken over by the Medical Research Council after the War, such as Perutz's work on the structure of haemoglobin. Gradually molecular biology began to form a significant part of the programme of the MRC, illustrated by the establishment of a lavishly endowed Laboratory of Molecular Biology in Cambridge in 1962. This was the year in which Max Perutz, John Kendrew, and Francis Crick (who were to work in the new lab along with Frederick Sanger and Hugh Huxley) shared Nobel prizes for work in molecular biology with James Watson then at Harvard, and Maurice Wilkins at King's College, London. The plans for the laboratory, drawn up by Perutz, had been enthusiastically received in 1958 by Sir Harold Himsworth, Secretary of the MRC from 1949 to 1969, and an influential supporter of structural studies in biology throughout this period.[90]

In his book on the organisation of research, Himsworth writes of knowledge not as a tree with divergent branches but as a sphere of ignorance, divided into 'natural provinces', from the surface of which, where practical problems are located, ultimately convergent lines of investigation are driven into the interior, and become increasingly esoteric, theoretical and general in scope.

> If a subject can be shown to be an integral part of a continuum, or a province of knowledge, which at its periphery engages directly with problems of evident social concern, then it is in no need of other justification . . . Considered in isolation molecular biology might seem (and indeed at one time did seem) to have scant claim for support for funds for medical research. Viewed however in the whole context of biomedical knowledge, it is at once evident that it is an indispensable constituent of understanding.[91]

One can infer that this was how molecular biology was viewed by the controlling figure at the MRC in the 1950s and 1960s, as part of a complex of projects linked up to clinical research, as a level of analysis beyond physiology and biochemistry, closer to the centre of the sphere of ignorance.

> As I see it, the conceptual basis of units of organisation for the further development of scientific knowledge is the province of natural knowledge, and the operational instrument corresponding to this is a central research organisation with nation-wide responsibilities . . . Policy in repect to the development of each of these provinces is entirely a matter for professional scientific judgement and the responsibility for this can only be vested in an essentially professional body.[92]

The task seems to reduce to selecting those individuals with the greatest potential as researchers, forming institutes around them and allowing them to press on into the interior. The role of the MRC is to select such investigators and to manage the interaction between different parts of the province. Whilst this scheme has a certain coherence and force, it seems to obscure or beg questions of priority, emphasis, strategy and the limitations on funding. As an ideology of practice, it naturalises the arrangements made by the MRC, and seeks to justify the autonomy it had sought to maintain from its formation in 1920.[93] In fact the MRC in the

1950s and 1960s seems to have placed considerable importance on molecular biology, with a large biophysics unit at King's College London and a Microbial Genetics Unit in London, which *later* formed the basis of the Department of Molecular Biology at Edinburgh, as well as the Cambridge lab, the third largest establishment run by the MRC and the only one of the three to be closely linked to a particular discipline. In 1968 the budget of the Cambridge lab was £503,000 and that of the King's College unit £273,000, out of an MRC expenditure of £6,967,000 on external staff, research units and other establishments, and a total budget of £16,653,000.[94] Drath has calculated that in 1968/9 some £7,000,000 was spent on biology at the molecular level, out of a total government expenditure on biological research of £70,000,000.[95] These figures are not very reliable, but one might bear in mind that the total budget of the National Institutes of Health in the same period was just over $1 billion. Indeed in the early 1960s when the Cambridge lab was being set up, the need for a European response in biology to *le défi Americain* was already being considered. In 1962 the Royal Society's ad hoc Biological Research Committee wrote of the need for 'major changes in the organisation of biology departments' if Britain were to keep its place in international research, a point of view not accepted and acted upon by the more traditionally-minded University Grants Committee.[96] In 1971 a Working Group on Biological Manpower of the Council for Scientific Policy noted:

> . . . we are convinced that in several of the more important areas of fundamental biology further advance demands interaction between biologists of different kinds, and between biologists and physical scientists, and that it is urgently necessary to establish a type of organisation which makes such interactions possible.[97]

Thus in governmental reviews of the state of biological research in Britain in the 1960s one can discern competing strategies, the more radical of which stressed the need for new skills, new modes of collaboration, new research institutions and a restructuring of disciplinary boundaries, in order to sustain the reductionist movement in biology. This was part of the context for plans drawn up in the early 1960s, with an initial stimulus from physicists connected with the nuclear physics centre at CERN in Geneva, to create a European Molecular Biology Organisation (EMBO). The intention was to provide training for biologists in the latest methods and techniques, to allow mobility within Europe, to set up meetings and workshops, to facilitate the establishment of new departments and to build a central laboratory. The idea of an EMBO was made public in 1964 by the organising group of molecular biologists, including Perutz, Kendrew and Monod. The Volkswagen Foundation soon stepped in to provide a large grant of 2.7 million DM($687,000) to pay for the workshops, conferences and collaborative visits for 1966, 1967 and 1968. This generosity has been attributed to the desire of a recently formed organisation to become involved with international science.[98] But one is also reminded of the Rockefeller Foundation, providing money to catalyse change on the technical, methodological, professional and institutional levels, to get scientists and governments to accept new forms of research. In this case the example of a highly specialised, technology-intensive, collaborative biology is being placed before European governments, as an instance of joint effort and a strategic move to stay with the competition by servicing biology from a central organisation. Interestingly the Volkswagen Foundation went on to provide substantial support for the Institute of Biotechnology in Brunswick, which now plays an important role in the large German programme of applied research in this area.

The EMBO plan of 1967 also envisaged a central laboratory costing $15 million to build and $8.5 million per year to run. In the discussions of 1968 and 1969 this idea was received very sceptically and the revised version of 1970 envisaged a less ambitious facility that would cost £2.5 million to build, with a stress on sophisticated instrumentation, employing about 60 professional scientists, most of them on short-term contracts. This lab was opened in Heidelberg in May 1978.[99] Even the second version of 1970 places a remarkable emphasis on expensive, complex research technology. Research projects were to be confined to those which were not likely to be carried out nationally because of being too expensive or requiring too large a group of scientists. There were to be two research themes: macromolecular assemblies and whole cell phenomena. The first of these was thought to require X-ray and neutron diffraction electron microscopy, NMR spectroscopy and the sequencing and synthesising of macromolecules, while the second required the growth of cell cultures, automatic separation of cell types, and rapid microanalytical methods. Clearly then the lab was to intervene in European biology on a considerable scale, by bringing in scientists to produce the technology for a new biology, and the biology for a new biotechnology. That tendency has been strengthened by the addition to the EMBO programme of a facility for rigorously contained recombinant DNA experimentation. The lab embodies a centralist, elitist technology and capital-intensive strategy for keeping up with the movement of highly specialised, rapidly progressing research; as such, it ran into criticism from other disciplinary groups, such as biochemists, and from less internationally minded research administrators.[100]

Finally a comparison of EMBO as a programme and an institution and the measures promoted by Weaver in the 1940s is illuminating. The Rockefeller programme in molecular biology was international in scope: funds were disbursed to various countries; scientists were encouraged to travel, to pick up the latest techniques and to explore new ways of working together; and moreover these moves were orchestrated from the Foundation, through the interaction of Warren Weaver, his fellow officials, the Trustees and their scientific advisory groups. By 1939, not only of course because of the Rockefeller example, this model of managed research was transferred to the state sector. The programme represented not only a re-classification and a directed reconstitution of biology, but also a stage in the incorporation of science, through the planned development of new forms of expertise in biology.

With important differences, EMBO seems to me a related exercise in the managed provision of new forms of expertise to service biology and biotechnology, through the development of a high-technology research institute and a network of very flexible, highly mobile, deeply competitive researchers, whose research programmes are now becoming increasingly articulated with those of the pharmaceutical, chemical and agricultural corporations. EMBO takes the form of an elite academy with state support, committed to turning research skills into technology, continuing to re-skill biologists at a high-level of specialisation, promoting changes of intensity and forms in national research programmes and to the central management and control of the programme of re-skilling through detailed assessment of who works in Heidelberg, who gets access to new techniques, and whose results can command further support. Furthermore this whole exercise in the development and restructuring of scientific labour is now being fitted together at the intergovernmental, institutional and individual levels with the formation of a biotechnology sector in European industry. Thus arises the 'directed autonomy' of a trans-national biology in the age of the multinational corporation as scientists plan, orchestrate and work on the restructuring and reconstitution of the labour processes of biology.

NATURE AS PROGRAM

In the first section I made some brief remarks on the emergence of biotechnology as a novel industrial and political phenomenon. I suggested that in order to understand the forces driving this development forwards we consider the activities of biologists — working with a specific conception of life, selecting out various problems, and building up new techniques, research technologies and forms of collaboration — as taking place in a condition of 'directed autonomy'. I also suggested that we think of molecular biologists as having developed a new language in which to formulate biological problems and corresponding forms of representing and 'making sense' of their practices as a new, more advanced form of biology, as work at the organising centre of the life sciences. In the second section I went on to suggest that molecular biologists have promoted a reductive shift in what counts as an explanation in biology, by concentrating attention on functions and processes considered in terms of molecular structure and interaction.

In the third section I described the changing institutional conditions of research in this century, with particular emphasis on the building up of a managed cadre of expertise in biology, through precedence being given to specific methods, styles of research and problem areas, by the encouragement of mobility between certain institutions, by the programmatic organisation of research and by the development of the techniques of research management for State-supported investigations. My discussion of EMBO suggests that the whole exercise could be seen as a system in which scientific skills are metabolised under the control of a network of laboratory directors linked up to a central facility. In other words, the skills of biological research (e.g. dissection, field obvservation, spotting mutants, doing biochemical assays) are continually being monitored, re-organised, refined, redistributed within groups of scientists, re-evaluated, broken down into more specialised tasks and regrouped within EMBO. There some skills are transformed into new technology, and others are diffused into a general programme of biology in Europe. In this sense one could describe the history of molecular biology as a process of re-skilling, or the planned reconstitution of biological labour power, in increasingly specialised forms.

As an example of the sort of specialisation I have in mind, one might quote the example of the young French molecular biologist Pierre Kourilsky — trained as a mathematician, now working on viral genetics — who has said that it was only when he was in hospital with hepatitis that he learnt where his liver was.[101] In other words he is able to function at the leading edge of this form of biological research, as are many others with similar backgrounds in the physical sciences, without needing to have a broad grounding in the life sciences, although such work is directly concerned with very general biological issues. Seemingly the molecular generalities of living organisms can best be apprehended by highly specialised technicians with minds totally uncluttered by knowledge of diversity. I have argued elsewhere that one of the interesting features of the highly schematic way in which molecular biologists conceive life is that it allows the ingress of physical scientists with very different notions of what living matter must be like.[102] Olby has shown, for example, that Francis Crick's work makes frequent use of a principle of simplicity (or natural economy) and derivations from it. His style as a theoretician, with its roots in physics, is embodied in statements about how nature must operate.[103]

But molecular biology is not only a dismissal of the natural historical tradition; it is also advance towards biology as the science of information. In the final section I want to draw out some more implications of the labour process perspective on molecular biology, by considering the significance of the integration of concepts like 'program', 'message' and 'code' into a reductionist biological framwork. What lies behind the idea of a genetic program? Again I should say stress that I am not saying that the idea that organisms are programmed is a biassed or deformed account of reality, a misconception occasioned by capitalism. What I am asking is why it should be that we attempt to apprehend nature as programmed? Why do we frame and analyse the manifold of nature in this way? What is it that leads us to this way of formulating questions?

I am not in a position to answer this query definitively. I propose, however, that we think about it by separating out three *mutually constitutive* levels of analysis, which represent three levels of process or types of determination — each depending upon, setting constraints for, and reacting back upon the others. The first concerns scientific concepts and ideas; the second, forms of scientific organisation; and the third, cycles of accumulation, forms of economic power and theories of society.

Scientific Concepts

The Second World War saw major developments in the field of computing, with the construction of several immense machines capable of performing mathematical calculations with unprecedented rapidity.

From wartime uses in ballistics, cryptography and bomb design, these machines began to be transferred to the scientific problems of peacetime, for example in astrophysics and X-ray crystallography. Thus in 1951 John Kendrew in Cambridge wrote the first computer program for analysing crystallographic data on protein structure.[104] But the novelty of these machines lay not only in their ability to handle enormous quantities of data and thereby to open up new problems for attack, but also in the fact that they required a program, a structured set of instructions, for their operation. What is interesting is that in due course this idea of a self-activating, self-controlling set of instructions found its way into biology and the human and social sciences. In doing so it transformed the concept of hereditary information.

In the 1930s and 1940s scientists were already asking the question of how molecular structures such as the protein (or DNA) components of chromosomes could embody hereditary information. How could biological specificity be packed into microscopic units such as sperm heads, bacteriophage or DNA molecules of known dimensions and of a certain conformational regularity?[105] Similarly those working on the genetics of bacteria and viruses came to think of their hereditary material as packets of information, or messages, that could be transferred into cells, or passed between them. But one can perceive a distinction between a message, information in transit, so to speak, and a program considered as an embodied set of instructions with complex properties of regulation and adaptation. Gradually the latter idea was able to assimilate the former, in a developed model of genetically controlled protein synthesis. The shift then is from a concern with 'passive', structural questions of encoding and the location of specificity , to more active issues of catalysis, control and the generation of form, the transitive aspects of the notion of program. The question is not so much where or how could information be packed, but rather how could programs be expressed, and how general is the applicability of this as a biological idea? The passages by Jacob quoted above suggest strongly, as does Watson's *The Molecular Biology of the Gene*, that some molecular biologists see the idea of a genetic program as a major organising principle in the life sciences, and that other questions and lines of research can be derived from focusing on it. To them, it makes sense to work from this level and this conceptual base.

Forms of Scientific Organisation

In the third section of this article I tried to to draw attention to the continual reconstitution of biology that has been going on since the beginning of this century, through the deliberate selection and development of new forms of expertise, new combinations and types of skills, new relations of status, power and control between disciplines and changes in the nature and structures of research support. I also argued that the institutions set up to fund major programmes of research have gradually been forced to develop systems of management and control, or an intensifying degree of oversight and direction over the disparate areas of investigation. In this sort of environment scientists have sought to provide rationales for particular patterns of research support, as bases for sets of research institutions or multifaceted investigation. In the case of biology such rationales or models of coherence are bound up with conceptions of life. Let us consider some examples. Donna Haraway, in her work on the dissenting, organicist tradition in twentieth century biology, writes of the work of the embyologist Paul Weiss.

> So life is process, not substance. It involves vast numbers of dynamic structures interacting in time and space; life consists of orderly, complex, group behaviour whose limits are defined by rules of order empirically unearthed. The idea of limits leads to that of levels.[106]

Compare this with the highly reductive formulation of the nature of life by Muller quoted above. Significantly Weiss' distinguished work on neural development placed little emphasis on genetics. In 1950 he was invited to set up a Division of Biology at the National Science Foundation, of which he wrote;

> I restructured the administrative subcategorisation of "biology",

> previously based on forms of life . . . or on methods of study . . . by a hierarchical system of order according to functional principles in common to living organisms; to list, in ascending order: Molecular, Cellular, Genetic, Developmental, Regulatory and Group and Environmental Biology.[107]

What was to become the leading agency in the United States for the support of fundamental biological research (leaving aside the role of the National Institutes of Health) was structured around this explicitly non-reductionist idea. Within that structure molecular biologists have lobbied vigorously for more resources and a greater degree of influence, leading up to the reorganisation of the late 1960s, which saw a major growth of interest in eukaryote cells and animal tumor viruses, and the establishment of a human cell biology programme, based on the idea that one type of cell should be exhaustively analysed. This emphasis on cell biology was designed to gain more power to define the terms and goals of basic biological research, being modelled on the ecologists' strategy within the International Biological Programme.[108] One can also regard Himsworth's remarks on 'natural provinces of knowledge' and the sphere of ignorance as a framework of ideas, with some sort of institutional correlative, within which programmes of research could be canvassed. In the case of the Laboratory of Molecular Biology in Cambridge and the Department of Molecular Biology in Edinburgh, that framework was supported somewhat against the wishes of academic supporters of existing disciplinary arrangements.

The point I am trying to make is that, in the contemporary research environment, if an area of research is to prosper and achieve influence, its proponents must be able to explain its possible role within a set of research programmes, a structured, 'rationalised' complex of research initiatives — or, to borrow a phrase from Michel Foucault, a 'regime of truth'.

> By "truth" is meant a system of ordered procedures for the production, regulation, distribution and circulation of statements . . . Truth is linked by a circular relation to systems of power which produce and sustain it, and to effects of power which it induces and which redirect it.[109]

I want to suggest therefore that we view the notion of a genetic program as an 'enabling concept', which facilitated the production of certain types of rhetoric and argument by molecular biologists, in their struggle to promote a specific form of reductionist programme within biology. Molecular biologists can claim to 'make sense' of other areas of biology by reference to this idea. In some cases they may not do so explicitly, but without a theoretical rationale for attempting to restructure biology their cause could not advance under present conditions. Just as the concept of the genetic program interacts with a developing structure of managed research, so that structure in turn is inscribed within a dynamic system of economic and political power, which has recently forced a shift by molecular biologists from the reprogramming of biology to the reprogramming of nature.

Cycles of Accumulation, Crisis and Ideology

When the idea of establishing a programme in molecular biology was first being considered, the economies of the Western World were deep in recession. In 1929 John D. Rockefeller bought a section of midtown Manhattan on which to build an International Center, to express his faith in the long-term future of capitalism once the contemporary disorders were removed.[110] In the mid-1930s the Foundation,based on the capital his companies controlled, began a series of moves to build up the institutions of research on a long-term basis — allowing scientists to travel, buying them the latest equipment, getting them to think programatically and developing forms of managerial control over their research.

By the end of the Second World War the economic and political order had been significantly altered and the condition of science and technology was much stronger. American capital had manoeuvred into a dominant position around the world, whilst the economies of leading European nations were in disarray caused by victory or defeat. The following decade saw the slow emergence of spheres of military politics and economic influence for the two leading world powers, America and the Soviet Union. Amongst other things the War had cleared away some of the obstacles to the process of accumulation. Despite the major perturbations of wars in the Middle East and South East Asia, the next decades saw a period of economic expansion for all the capitalist countries, albeit at markedly different rates. The expansion was sustained by high levels of state expenditure, particularly on welfare provision and health care, on military and related technology, and on the indirect subsidy of privately owned industrial corporations. Within each of the major industrial economies some firms were able to push the process of growth and economic concentration to unprecedented lengths, and to co-ordinate their production and sales on a multinational basis. They also exerted a growing influence in national and international politics towards the survival and development of the capitalist system itself, at the expense, if need be, of national or sectoral interests. Both nationally and internationally, political choice between parties and policies became increasingly restricted to alternative programmes of system management, setting modernisation against tradition and effacing the plausibility of radical social change.

Moreover social analysis came increasingly to adopt the reifying perspective of the systems approach, for example in operational research, functionalist sociology, cybernetics, organisation theory, ergonomics, time and motion study, applied psychology. At the same time the planning, production and marketing activities of industrial firms, and the activities of growing numbers of state institutions, concerned with some aspect of social or economic management, have become increasingly dependent on a cadre of technicians in the natural and social sciences, concerned with analysing, simulating, monitoring and modifying social units apprehended as more or less smoothly functioning systems. Warren Weaver himself linked together different types of research in a reifying schema, in an article written in the late 1930s but published in 1949, at a time when the fields of operations research and systems theory, with which he had been concerned in the war, were being taken up by industry.

> Is a virus a living organism? What is a gene, and how does the original genetic constitution of a living organism express itself in the developed characteristics of the adult? Do complex protein molecules "know how" to reduplicate their pattern and is this an essential clue to the reproduction of living creatures? All these are certainly complex problems, but they are not problems of disorganised complexity, to which statistical methods hold the key. They are all problems which involve dealing simultaneously with a sizeable number of factors which are interrelated into an orgainc whole. They are all, in the language here proposed, problems of organised complexity.
>
> On what does the price of wheat depend..? This too is a problem of organised complexity... How can currency be wisely and effectively stabilised? To what extent is it safe to depend on the free interplay of such economic forces as supply and demand? To what extent must systems of economic control be employed to prevent the wide swings from prosperity to depression? These are

> also complex problems, and they too involve analysing systems which are organic wholes, with their parts in close interrelation.
>
> How can one explain the behaviour pattern of an organised group of persons such as a labour union, or a group of manufacturers, or a racial minority? . . . With a given total of national resources that can be brought to bear, what tactics and strategy will most promptly win a war, or better: what sacrifices of present selfish interest will most effectively contribute to a stable, decent and peaceful world?[111]

Thus one of the early managers of molecular biology describes how he thinks the social analysts of the post-War era should think. The ideology of practice he proposes represents the social and biological sciences as the analysis of specific types or manifestations of complexity. A systems biology extends the systems perspective down to the molecular level. Again perhaps I should make it clear that I am not saying that the systems approach does not 'work' scientifically or that it is a distortion of what is 'really going on in living organisms'. Clearly it does work, under specific criteria of validity. But given that it seems to be a serviceable account of how organisms and societies operate, we should also ask: what are the consequences of thinking of these entities as mere instances of complexity? How does it affect our understanding of the research if we take bacteria to be programmed complex systems, somewhat like other complex entities, similarly conceived in systems terms, like factories or business enterprises?

I suggest that one effect is to diminish scientists' understanding of the social basis of their commitment to particular forms of thought or types of explanations, so that they become more easily moved around by peer groups, sponsors and external agencies from one instance of 'complexity' to another. A reified nature loses many of its former implicit meanings and offers few handholds to those who wish to resist the continual rebuilding of science, by virtue of focusing on intrinsic divisions, boundaries or properties. Nature is mere complexity, and life a term with no real meaning.

Another effect is to naturalise the systems approach to social analysis. It becomes 'natural' to think of, say, a factory as an organic system like a living cell — taking in raw materials, dependent on sources of energy, building up new components and exporting some of them, and maintaining itself in some environment, under the control of a central plan for the cell, the genetic program.

I have also tried to suggest that the history of molecular biology in the 1950s and 1960s is the history of the organisation of new forms of biological skill and expertise. These were organised within a reductive programme of research, and directed by a set of communicating, national scientific elites who were able to retain a degree of autonomy over the formulation and execution of research. With the revaluation of the principles and goals of State-supported research, catalysed by deepening economic recession since the early 1970s, the institutions, programmes and practitioners of molecular biology have been moved into a phase of more rapid and determined incorporation. As described in the introduction, this was achieved through a slight reorientation of the already developed relations of control between industrial capital, government research planning agencies, directors of academic research and their resources of living and dead biological labour.

CAPITALISING LIFE

I want finally to consider one aspect of this process, the production of reprogrammed micro-organisms and the patenting of 'new' forms of life, as a response to an economic challenge. In the early 1970s, several technical advances in biochemistry and microbial genetics came together which seemed to indicate that a much greater degree of precision and control would be possible in transferring genetic elements between cells , for example from relatively complex and poorly understood eukaryotic cells to bacteria. In other words single elements from a genetic program could be incorporated into simpler organic systems to discover how those instructions were expressed, or to reprogram the bacterium (or other host cell) with new instructions to make molecules synthesised only by higher organisms. A topical example is human insulin, made only in human pancreatic cells, although most diabetics can use calf or pig insulin. A number of research groups in the United States have recently succeeded in transferring functional insulin precursor genes into bacterial cells and getting usable quantities of human insulin from this process. Other groups are working on interferon, others on the bacterial production of medically significant hormones and proteins (such as clotting factor 8), others on the development of new vaccines by the excision at the molecular level of those parts of viruses which are harmful from those parts which will cause antibody production and confer immunity. Many of these goals offer very significant profits through the massive use of these therapeutic agents within hospital-based systems of medical care. The market for insulin in the United States is about $180 million annually, of which around 80% is already controlled by Ely Lilly, the company most heavily involved, through several parallel contracts, in the bacterial production of insulin. Other pharmaceutical corporations like Hoffman La Roche, Glaxo, Burroughs-Wellcome and Hoechst have also been active in this field for several years, monitoring the latest developments, bringing in consultants and new employees from academic life, starting new research groups, and placing contracts with the research companies (like Biogen and Genentech) to keep up with the competition.

One consequence of this rapid increase in industrial interest has been a dramatic shift in the social and legal conventions surrounding the the property rights to the fruits of academic research. Individual researchers have withdrawn from communication with colleagues, even in the same lab; patent applications for new organisms and techniques have been filed, by universities, individuals and companies, sometimes with disputes over ownership. Similarly items of information, such as the base sequences of particularly important genes, have acquired a real, if temporary, cash value. But not only have claims been laid to bits of programmed nature — transitory and tactical claims as well as those intended to last for the life of a patent — a debate has also sprung up over whether living nature can be commoditised and owned. This debate has in turn forced a redefinition of life, so that 'new' life forms can be privately owned. It is not simply that 'life is chemical' as the Monsanto advert claims, but that life can be made in chemical forms that can be classified as 'unnatural' in origin, and hence ownable artefacts, whilst also being part of living nature.[112]

Similarly recombinant DNA techniques are being applied in agriculture, food production and plant breeding to produce new strains of plant, new forms of food and new types of fertiliser. One feature of this process is the recent purchase of a number of seed firms by oil companies attempting to move into new areas as the supply of oil, from which fertilisers are made, falls. The

US SUP. COURT
TAXPAYERS
NSF & NIH
PRIVATE CORPORATIONS
CONSUMERS
MARKETING MICROORGANISMS - THE GREEN HOLE

strategy seems to be to develop new plant strains, less dependent on chemical fertilisers. At the same time national and international organisations are making moves to protect the world's diminishing resources of plant types, as tropical forests are destroyed, by establishing a World Seed Bank and a Plant Gene Bank, on which plant breeders will be able to draw. It is almost certain that Third World plant genes will be incorporated into new plant strains, to be sold back to those countries at high prices, and in ways that sustain their technological and economic dependence.

Who then owns plant genes and what are the rights of ownership? In this sense 'biotechnology' is not simply a way of using living things that can be traced back to the Neolithic origins of fermentation and agriculture.[113] As technology controlled by capital, it is a specific mode of the appropriation of living nature — literally capitalising life. It constitutes a new set of social relations between fundamental research scientists, industrial corporations and state agencies. Its *telos* is to promote technology while regulating the deskilling and displacement of workers. It also serves to inhibit collective resistance on the part of 'consumers' of health services, expropriated peasants and custodians of plant gene resources.

Biotechnology — formed from the institutions, skills and concepts of a directed molecular biology — sustains the circulation and accumulation of capital by offering a new set of products, processes and markets. Life becomes a productive force, playing a growing role in reconstituting the social relations of a new capitalist order.

Notes

1. 'Biotechnology Back in the Limelight', editorial, *Nature, 283* (10 January 1980), 119; see also the articles in this issue on pp.122-131.
2. 'Curing Europe's Biological Death-Wish' editorial, *Nature, 283* (24 January 1980), 319; see also the articles in this issue on pp.323-5, 328-9; also 'Eleventh Hour for Biotechnology in Britain', editorial, *Nature, 284* (10 April 1980), 499.
3. DNA (deoxyribonucleic acid) is the substance from which the genes of nearly all organisms are made. Recently it has become possible to excise specific sections of DNA molecules, that is, to extract specific genes or gene fragments from living cells, to transfer them into other organisms and get the biological information, which is encoded in the excised genes, expressed in the host cells. This is a process of controlled recombination of sets of genetic instructions; the resulting molecules are therefore called recombinant DNA molecules. When a new set of instructions is added to a bacterium, we can think of that cell as having been 'reprogrammed'. In the text, however, I also ask the question of what it means to think about cells and organisms in this way.
4. Interferon is the name for a group of substances produced by mammalian cells in an attempt to resist viral attack. See P. Newmark, 'Engineered *E. Coli* Produce Interferon', *Nature, 283* (24 January 1980), 323; F. Balkwill, 'What Future for the Interferons', *New Scientist* (24 January 1980), 230-232.
5. At current prices, one pound of interferon would cost $22 billion — if there were that much of it in existence. Other examples include insulin, cortisone and proteins which facilitate blood clotting. Similar technological strategies are being opened up in agriculture and food production. For example, Tate and Lyle, having failed in West Africa to harvest successfully berries which produce a powerful natural sweetener, talin, are now planning to make this substance synthetically using reprogrammed yeast cells. This venture brings together Tate and Lyle Ltd., the Wolfson Foundation and the Biological Laboratory, Kent University. See 'Powerful Sweetener Sought from Microbe', *Financial Times* (16 April 1980), 6.
6. See 'Industry Starts To Do Biology With Its Eyes Open', *The Economist* (2 December 1978), 95-102; David Dickson, 'Private Actions Raise Public Eyebrows', *Nature, 278* (5 April 1979), 494-5; Pearce Wright, 'Time for Bug Valley?' *New Scientist, 83* (5 July 1979), 27-29; J. Wilkinson, 'Engineering a Genetic Revolution', *New Scientist, 85* (6 March 1980), 728-9; R. Walgate, '600 Apply to Biogen', *Nature, 284* (17 April 1980), 585.

7. *Nature* editorial, op.cit. (note 2). See also R. Walgate,'Biotechnology Report Urges £10 million Programme to Match Competitors', *Nature, 283* (24 January 1980), 324-5; R. Walgate, 'R and D Attitudes Wrong, Says Report', *Nature, 284* (10 April 1980), 502.
8. See B.J.M. Miflin, 'Breeding More Protein', *Nature, 275*(12 October 1978), 483-4; (this article describes the breeding programme for cereals and legumes of the Food and Agriculture Organisation [FAO] and the International Atomic Energy Authority [IAEA], which now includes genetic manipulation experiments). G. Harvey, 'Green Gene Manipulation', *New Scientist, 77* (23 March 1978), 782; (this article describes the allocation of £300,000 in additional funds by the Department of Education and Science [DES] in the UK to the Agricultural Research Council [ARC] for a programme of genetic engineering work linking 4 plant breeding institutes). R. Riley, R.B. Flavell, 'Recombinant DNA and Agricultural Research', in J. Morgan, W.J. Whelan, (eds.), *Recombinant DNA and Genetic Experimentation* (Oxford: Pergamon Press, 1979). Ralph Riley, formerly Director of the Plant Breeding Institutes in Cambridge and a member of the influential Ashby Committee on genetic manipulation, is now secretary of the ARC. The former Secretary, Sir William Henderson, also a member of Lord Ashby's committee in 1974, is now Chairman of the Genetic Manipulation Advisory Group (GMAG), which controls the safety of genetic engineering in the UK.
9. For example ICI has given money to the University of Leicester to establish a new laboratory devoted to industrial recombinant DNA research.
10. S.Krimsky, D. Baltimore, 'The Ties That Bind Or Benefit?', *Nature, 283* (10 January 1980), 130-131. David Baltimore, a Nobel prizewinner and a member of the advisory body on the safety of recombinant DNA research in the US, has commercial interests in this field.
11. Articles by people critical of these developments in recombinant DNA research include: L.Cavalieri, 'New Strains Of Life Or Death?' *New York Times* magazine (22 August 1976), 8 et seq.; J.King, 'A Science for the People', *New Scientist, 74* (16 June 1977), 634-5; J.King, speech at ASTMS meeting on 'Genetic Engineering, Its Safety and Social Implications', London, October 1978, reported in *Medical World* (November/December 1978), 9-15; J. King, transcribed interview (30 August 1979), Documents submitted to Governmental hearings on Genetic Engineering Law, September 1979 available from Der Bundesminister für Forschung und Technologie, Postfach 20 07 06, 5300 Bonn 2, Federal Republic of Germany; G. Wald, transcribed

interview from the same source; J. Beckwith, 'Recombinant DNA; Does The Fault Lie With Our Genes?' *Science for the People* (May-June 1977), 14-17; Genetics and Social Policy Group of Science for the People, *The Health Hazards of Gene Implantation* (Boston, Mass: pamphlet undated); N.M Teich, R.A. Weiss, 'Beware The Lurking Virogene' in R.F. Beers, E.G. Basset (eds.), *Recombinant Molecules Impact on Science and Society* (New York: Raven Press, 1977), 471-483; British Society for Social Responsibility in Science (BSSRS) Genetic Engineering Group, *Genetics, Medicine and Social Control,* in *Report of the House of Commons Select Committee on Science and Technology* (London: HMSO, Sessional Papers 1978-9); E. Yoxen, 'The History, Social Relations and Possibilities of Molecular Biology', *Science for People* 43/44 (1979), 16-19; E.Yoxen, 'Regulating The Exploitation Of Recombinant Genetics, in Ron Johnston, Philip Gummett, (eds.), *Directing Technology: Policies for Promotion and Control* (London: Croom Helm 1979), 225-244; E.Yoxen, Review of June Goodfield *Playing God: Genetic Engineering And The Manipulation Of Life*, (London: Hutchinson, 1977), *Radical Science Journal 10* (1980); E.J. Yoxen, 'Recombinant DNA Research And Alternative Biotechnologies', in *Neucombinationen von Genen* (Frankfurt: Batelle-Institut e.v.; 1979), 244-250 (this volume also contains a number of critical statements, in German from Herbig, Hohfeld and Altner, and in English, from Chargaff and Signer); E. Chargaff, 'Uncertainties Great, Is The Gain Worth The Risk?', *Chemical and Engineering News* (30 May 1977), 32-35; S. Wright, 'The Recombinant DNA Advisory Committee', *Environment, 21, 3* (April 1979), 2-5; S. Wright, "DNA: Let the public choose", *Nature (12 October 1978), 468; S. Wright, 'Setting Science Policy: The Case Of Recombinant DNA', Environment, 20* (May 1978) 6-15, 39-40; S. Wright, 'Doubts Over Genetic Engineering Controls', *New Scientist, 72* (2 December 1976), 520; F.E. Simring, 'Recombinant DNA: Technological Innovation vs. Environmental Protection', *Chemistry and Industry,6* (18 March 1978), 174-5; T.Howard, J. Rifkin, *Who Should Play God?* (New York: Dell, 1977); J. Rifkin, 'DNA: Have The Corporations Already Grabbed Control Of New Life Forms?', *Mother Jones* magazine (February-March 1977), 23-26,39; C.Manwell, C.M.A. Baker, *Genetic Engineering: Risk, Politics And Dissident Scientists* (Manuscript, 1979); Agata Mendel (pseudonym), *Les Manipulations Génétiques* (Paris: Editions du Seuil, 1980); Jost Herbig, *Die Gen-Ingenieure* (Munich: Carl Hanser Verlag, 1978).

12. These ideas are discussed in E.J. Yoxen, *The Social Impact of Molecular Biology* (Unpubl. Ph.D. Thesis, University of Cambridge, 1978).

13. For a statement of this view see S. Brenner, 'New Directions In Molecular Biology', *Nature, 248* (26 April 1974), 785-7.
14. Michel Foucault, *The Order of Things: An Archaeology of the Human Sciences* (1966; London: Tavistock, 1970), p. 160.
15. Ibid, p. 162.
16. A.I. Oparin, *Life, its Nature, Origin and Development* (1960; London: Oliver and Boyd, 1961).
17. A.I. Oparin, 'The Problems of the Origin of Life', *Modern Quarterly, 6,* 2 (1951), 135-146.
18. H.J. Muller, 'The Gene as the Basis of Life' *Proceedings of the International Congress of Plant Science, 1* (1926), 897-921; reprinted in H.J. Muller, *Studies in Genetics* (Bloomington: University of Indiana Press, 1964), 188-204; quotation at pp. 199-200.
19. Ibid, p. 201.
20. E.A. Carlson, 'An Unacknowledged Founding of Molecular Biology: H.J. Muller's Contribution to Gene Theory, 1910-1936', *Journal of the History of Biology, 4* (1971), 149-170; N. Roll-Hansen, 'Drosophila Genetics: A reductionist research program', *Journal of the History of Biology, 11* (1978), 159-210.
21. J.D. Bernal, "Definitions of Life", *New Scientist, 33* (5 January 1967), 12-14, at p. 13; this article is an extract from *The Origin of Life* (London: Weidenfeld and Nicolson, 1967).
22. Ibid, p.14.
23. F. Jacob, *The Logic of Living Systems* (1970; London: Allen Lane, 1974).
24. This view of molecular biology is not to be found in any of the existing historical studies of molecular biology. See R.C. Olby, *The Path to the Double Helix* (London: Macmillan, 1974); F.H. Portugal, J.S. Cohen, *A Century of DNA: A History of the Structure and Function of the Genetic Substance* (London: MIT Press, 1977); H.F. Judson, *The Eighth Day of Creation: The Makers of the Revolution in Biology* (London: Jonathan Cape, 1979). For comments on these works, see E.J. Yoxen, 'The History of Molecular Biology' *British Journal in the History of Science, 11,* 39 (1978), 278-9; E.J. Yoxen, 'Molecular Biologists and their World', *Science, 205* (14 September 1979), 1119-20.
25. M. Foucault, 'Truth and Power Interview,' translated by F. Edholm, *Critique of Anthropology, 4,* 13/14 (Summer 1979), 131-7, at p. 134.
26. See for example H. Rose, S. Rose, 'The Incorporation of Science', in H. Rose, S. Rose, (eds.), *The Political Economy of Science* (London: Macmillan, 1976), 14-31.
27. See Raymond Williams, *Marxism and Literature* (London: Oxford University Press, 1977), especially Ch. 8; Raymond Williams, in 'The Role of the Mass Media: a discussion', *Wedge* 1 (1977), 33-37.

28. An alternative approach to the phenomenon of incorporation is to be found in the work of various German writers on science who use the concept of 'finalisation', with which they try to describe the ways in which specific goals are built into scientific models and theories, so that the resulting programme of research moves in a quasi-autonomous way towards specific applications. See G. Bohme, W. Van den Daele, W. Krohn, 'Finalisation in Science', *Social Science Information, 15,* 2/3 (1976), 307-330; G. Bohme, W. Van den Daele, W. Krohn, 'The "Scientification" of Technology', *Sociology of the Sciences, 2* (1978), 219-250; K. Buchholz, 'Verfahrenstechnik (Chemical engineering) — its Development, Present State and Structure', *Social Studies of Science, 9,* 1 (1979), 33-62; R. Hohlfeld, 'Theory Development in Molecular Biology" (Paper presented at conference on *Theory of Knowledge and Science Policy*, Ghent, Belgium, December 1979).
29. S. Pickvance, ' "Life" in a Molecular Biology Lab', *Radical Science Journal* 4 (1976), 11-28.
30. L. Hodgkin, 'Politics and Physical Sciences', ibid, 29-60, at p. 43.
31. S. Hall, 'Culture, the Media and the "Ideological Effect" ', in J. Curran, M. Gurevitch, J. Woollacott (eds.), *Mass Communication and Society* (London: Edward Arnold, 1977), 315-348, at p. 322.
32. Ibid, 330.
33. M.J. Mulkay, 'Norms and Ideology in Science', *Social Science Information, 15*, 4/5 (1976), 637-656.
34. See Olby, op. cit. (note 24).
35. R.M. Young, 'Why Are Figures So Significant? The role and critique of quantfication', in J. Irvine, I. Miles, J. Evans (eds.), *Demystifying Social Statistics* (London: Pluto Press, 1979), 63-74; see also R.M. Young, 'Evolutionary Biology and Ideology: then and now', *Science Studies, 1* (1971), 177-204.
36 Jacob, op. cit. (note 23), p. 16.
37. Ibid, p. 13.
38. See also Garland Allen, *Life Science in the Twentieth Century* (Cambridge: Cambridge University Press, 1978).
39. See Roll-Hansen, op. cit. (note 20), on Morgan as an empiricist filter or obstacle to the more theoretical initiatives of geneticists like Muller. The same ideas appear in G.E. Allen, *Thomas Hunt Morgan: the Man and His Science* (Princeton: Princeton University Press, 1978).
40 See Olby, op. cit. (note 24); Judson, op. cit. (note 24). On Bernal's excitement of the potential of his diffraction pictures from crystalline pepsin, see D.C. Hodgkin, D.P. Riley, 'Some Ancient History of Protein X-Ray Analysis', in A. Rich, N. Davidson (eds.), *Structural Chemistry and Molecular Biology* (San Francisco: Freeman, 1966), 15-28.

41 See R. Lewontin, 'The Bases of Conflict in Biological Explanation', *Journal of the History of Biology, 2* (1969), 35-53.
42 The work of the molecular biologist, Francis Crick, is often given as an example here. See R.C. Olby, 'Francis Crick, DNA and the central dogma', *Daedalus, 99* (Fall 1970), 938-987.
43 The effect of this institutional context on Avery's research is clearly brought out in R. Dubos, *The Professor, the Institute and DNA* (New York: Rockefeller University Press, 1976).
44 Jacob, op. cit. (note 23), p. 8.
45 Olby, op. cit. (note 42).
46 Judson, op. cit. (note 24), esp. ch. 6.
47 Ibid, p. 488.
48 J.D. Watson, *The Molecular Biology of the Gene* (San Francisco: Freeman, 1965; revised edition, 1970; 3rd edition, 1976).
49 See for example J. O'Connor, *The Corporations and the State: Essays in the Theory of Capitalism and Imperialism* (New York: Harper and Row, 1974).
50 I. Gough, *The Political Economy of the Welfare State* (London: Macmillan, 1979), p. 66. For further discussions on the theory of the state relevant to science, technology and medicine see V. Navarro, 'Social Class, Political Power and the State: their implications in medicine' in *Medicine Under Capitalism* (London: Croom Helm, 1979), 183-229; L. Doyal & I. Pennell, *The Political Economy of Health* (London: Pluto Press, 1980).
51 A useful introduction to this area is C. Freeman, *The Economics of Industrial Innovation* (Harmondsworth: Penguin, 1974).
52 S. Coben, 'American Foundations as Patrons of Science: the commitment to individual research' in N. Reingold, *The Sciences in the American Context: New Perspectives* (Washington: Smithsonian Institution Press, 1979), 229-248.
53 M. Heirich, 'Why We Avoid the Key Questions: How Shifts in Funding of Scientific Inquiries Affect Decision-Making About Science', in S. Stich & D. Jackson (eds.), *The Recombinant DNA Debate* (Ann Arbor: Univ. of Michigan Press, 1977), pp. 234-260.
54 Robert E. Kohler, 'The Management of Science: The Experience of Warren Weaver and the Rockefeller Foundation Programme in Molecular Biology', *Minerva, 14* (Autumn 1976), 279-306, at p. 282. See also E.R. Brown, 'He Who Pays the Piper: Foundations, the Medical Profession and Medical Education', in Susan Reverby and David Rosner (eds.), *Health Care in America: Essays in Social History* (Philadelphia: Temple Univ. Press, 1979), pp. 132-153; E.R. Brown, *Rockefeller Medicine Men: Medicine and Capitalism in America* (London: University of California Press, 1979).

55 S.P. Strickland, *Politics, Science and Dread Disease: a Short History of US Medical Research Policy.* (Cambridge: Harvard University Press, 1972), p. 6.

56 R.E. Kohler, 'Walter Fletcher, F.G. Hopkins and the Dunn Institute of Biochemistry: a case study in the patronage of science', *Isis, 69* (September 1978), 331-355.

57 R.E. Kohler, 'Warren Weaver and the Rockefeller Foundation Program in Molecular Biology: a case study in the management of science', in Reingold, op. cit. (note 52), pp. 249-294. This is a revised version of an article in *Minerva, 14* (1976), 279-306.

58 See Allen, op. cit. (note 39).

59 See also G.E. Allen, 'T.H. Morgan and the Emergence of a New American Biology', *Quarterly Review of Biology, 44* (June 1969), 168-88.

60. T.H. Morgan, "A New Division of Biology", *California Institute of Technology Bulletin* (1928), 16-17; quoted in G.E. Allen, op. cit. (note 39), p. 345.

61. Kohler, op. cit. (note 57).

62. C.W. Pursell,' The Anatomy of a Failure: the Science Advisory Board, 1932-1935', *Proceedings of the American Philosophical Society, 109* (December 1965), 342-51. A recent work, which deals with the activities of leading physicists in America is D.J. Kevles, *The Physicists: the History of a Scientific Community in Modern America* (New York: Random House, 1979). This book includes a massive bibliographic essay on the social relations of physics, including a great deal of material on the 1920s and 1930s.

63. K. Birr, 'Industrial research laboratories', in Reingold, op. cit. (note 52), pp. 193-208.

64. Strickland, op. cit. (note 55).

65. These biographical remarks are taken from the transcribed oral history memoir provided by Warren Weaver in the Oral History Record Office (Record No. 434, 3 vols., 1961) at Columbia University, New York. I am grateful to Charles Weiner for bringing it to my attention. These volumes are referred to hereinafter as Weaver. See also W. Weaver, *Scene of Change: a Lifetime in American Science* (New York: Scribner's, 1970).

66. Weaver, 333-4.

67. Weaver, 502-3.

68. Weaver, 430, 432.

69. Weaver, 404-5.

70. Weaver, 401-2.

71. See Kohler, op. cit. (note 57), p. 274.

72. This point was made to me by Seymour Cohen.

73. *Rockefeller Foundation Annual Report*, 1938, pp. 39-40, 203-217, at p. 203.

74. C. Pursell, 'Science Agencies in World War II: The OSRD and its challenges', in Reingold, op. cit. (note 52). See also Kevles, op. cit. (note 62), esp. chapters 8, 9 and 10, and the bibliography where other studies of the OSRD are mentioned.
75. Strickland, op. cit. (note 55), p. 15.
76. Ibid, p. 18.
77. Ibid, p. 25-6.
78. C.V. Kidd, *Oral History Memoir* (Health Sciences Project, Columbia University), pp. 39-40.
79. The growth of the National Institutes of Health is described in Strickland, op. cit. (note 55), and in M.D. Reagan, *Science and the Federal Patron* (New York: Oxford University Press, 1969).
80. J.A. Shannon, *Oral History Memoir* (Health Sciences Project, Columbia University), p. 17.
81. Ibid, p. 20.
82. Ibid, p. 53.
83. Strickland, op. cit. (note 55); Reagan, op. cit. (note 79); R.A. Rettig, *Cancer Crusade: the Story of the National Cancer Act, 1971* (Princeton: Princeton University Press, 1977).
84. Strickland describes in detail the American Medical Association opposition to the Democrats' post-War National Health Program. which was to include federal aid for medical education. He writes, 'one of the messages received during the battle of 1951 was that the AMA would at least not oppose the Congress if it moved to open up new fronts in the war it still wanted to wage, through research, against dread disease'. (Op. cit. p. 74.)
85. J.S. Murtaugh, 'Biomedical sciences', in J.A. Shannon (ed.), *Science and the Evolution of Public Policy* (New York: Rockefeller University Press, 1973), 155-189; Heirich, op. cit. (note 53 above); R. Stevens, *American Medicine and the Public Interest* (New Haven: Yale University Press, 1971).
86. See Pickvance, op. cit. (note 29).
87. As Robert Walgate reports (see note 6), 600 young scientists have already applied to work in the new labs of Biogen in Geneva.
88. C.H. Waddington, 'Some European Contributions to the Prehistory of Molecular Biology", *Nature, 221* (25 January 1969), 19-22.
89. J.G. Crowther, *Science in Liberated Europe* (London: Pilot Press, 1949).
90. M.F. Perutz, 'Origins of Molecular Biology', *New Scientist, 85* (31 January 1980), 326-9; see also A. Landsborough-Thomson, *Half a Century of Medical Research: the Programme of the Medical Research Council, 2* (London: HMSO, 1975).
91. H. Himsworth, *The Development and Organisation of Scientific Knowledge* (London: Heinemann, 1970), p. 58.
92. Ibid, pp. 140-1.

93. Kohler, op. cit. (note 56); Landsborough-Thomson, op. cit. (note 90).
94. These figures are taken from L. Drath, *European Co-operation in Molecular Biology* (Unpublished Ph.D. thesis, University of Manchester, 1973); see Table 4, 3-26.
95. Ibid, pp. 3-21, 3-23.
96. *Report of the Ad Hoc Biological Research Committee*, Advisory Council for Scientific Policy Report, 1961-2 Appendix B, (London: HMSO, Cmnd 1920, 1962), pp. 17-31. Drath writes, 'The Advisory Council for Scientific Policy also accepted that money for the required changes would not be forthcoming from the University Grants Committee and advocated, as an alternative, the building up of a few selected centres of research.' Drath, op. cit., pp. 3-9.
97. Council for Science Policy, *Report of the Working Group on Biological Manpower* (London: HMSO, Cmnd, 4737, 1971), p. 37.
98. Sir John Kendrew, sometime Secretary-General of EMBO, quoted in Drath, op. cit. (note 94), pp. 5-7.
99. P. Newmark, 'European molecular biology', *Nature, 273,* (18 May 1978), 183.
100. Anon., 'The EMBO question debated', *Nature, 224* (1 November 1969), 406-8.
101. P. Kourilsky, transcribed interview (1976), Recombinant DNA Oral History Archive, Massachusetts Institute of Technology Library, p. 3.
102. E.J. Yoxen, 'Where Does Schrödinger's *What is Life?* Belong in the History of Molecular Biology?', *History of Science, 17* (1979), 17-52.
103. Olby, op. cit. (note 42).
104. Judson, op. cit. (note 24), p. 564.
105. See for example the discussions between Crick and Delbrück reported in R.V. Jones, *Most Secret War*, Hodder & Stoughton, 1978.
106. D. Haraway, *Crystals, Fabrics and Fields: Metaphors of Organicism in Twentieth Century Developmental Biology* (London: Yale University Press, 1976), p. 185.
107. Ibid, 367-8.
108. Transcribed interview with Herman Lewis of the National Science Foundation, Recombinant DNA Oral History Archive, MIT, 1975.
109. Michel Foucault, 'The Political Function of the Intellectual', *Radical Philosophy* 17 (1978), 14-17, at p. 14.
110. P. Collier, D. Horowitz, *The Rockefellers: an American Dynasty* (New York: New American Library, 1976).

111. W. Weaver, 'Science and Complexity', *American Scientist, 36* (1948), 536-544, at pp. 539-540.
112. 'Surely the line between life and non-life is so fine as to baffle even the experts in the art and at times cannot be drawn with conviction; compositions of matter of manufactures near the periphery cannot conveniently be deemed patentable or unpatentable on so ephemeral a ground.' So claims an Amicus Curiae (friend of the court) brief filed by the University of California at Berkeley. Quoted from D. Dickson, 'Patenting Living Organisms — how to beat the bug-rustlers', *Nature, 283* (10 January 1980), 128-9; see also C. Cookson, 'Supreme Court Considers Key Case of Genetic Engineering', *Times Higher Education Supplement* (11 April 1980), 4.
113. For an interesting discussion of what biotechnology connotes and of how the disciplines and problem areas involved might be directed and controlled by governments see A. Rip, Walter van der Es, 'Biotechnology: its Development and Context', a translation of an article to appear in *Wetenschapsontwikkeling en Wetenschapssturing* (Alphen a/d Rijn: Sansom, 1980).

Grunwick:
The Social Contract Meets the 20th Century Sweatshop

LES LEVIDOW

Introduction

We tend to think of modern technology as eliminating the autocratic management practices of the 19th century sweatshop, which exploited particularly vulnerable workers in a labour-intensive production process. However, the defeated strike at the Grunwick photo processing plant should make us think again, for the strike support strategy which led to defeat tended to treat Grunwick's anti-union stance as an obsolete vestige of a bygone era.

Yet in reality Grunwick had been incorporating some of the most advanced photoprocessing and computer technology, geared to carving out its particular consumer market in the modern leisure industry. Because Grunwick could project a modern, clean image of itself in the media, it was well suited as a test case for the anti-union politics being developed by the reactionary National Association for Freedom (NAFF).

With the May 1979 election, that politics has since progressed from a Tory pressure group to the ruling party in Britain, posing new challenges for any revolutionary strategy. It is important, then, for us to understand how the anti-union offensive which began at Grunwick has arisen not as some aberration but as a concomitant of modern technology.

Furthermore, the understanding presented here was worked out only after the Grunwick strike, through discussion with other comrades involved in it. As such a discussion was not possible during the months when many people were mobilised around the Grunwick strike demands, we will need in future to find ways and means of raising such questions in the actual course

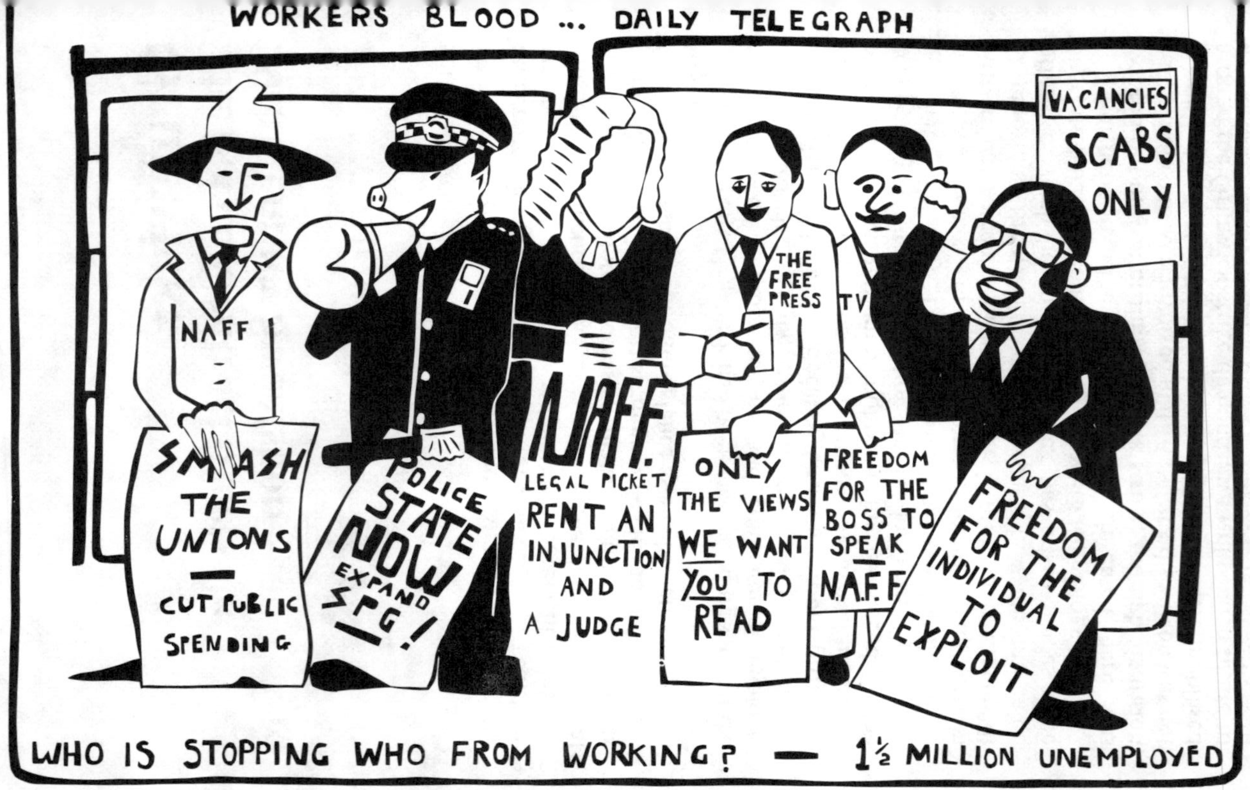
WORKERS BLOOD ... DAILY TELEGRAPH
NAFF
SMASH THE UNIONS — CUT PUBLIC SPENDING
POLICE STATE NOW EXPAND SPG!
NAFF. LEGAL PICKET RENT AN INJUNCTION AND A JUDGE
THE FREE PRESS
ONLY THE VIEWS WE WANT YOU TO READ
TV
FREEDOM FOR THE BOSS TO SPEAK N.A.F.F
FREEDOM FOR THE INDIVIDUAL TO EXPLOIT
VACANCIES
SCABS ONLY
WHO IS STOPPING WHO FROM WORKING? — 1½ MILLION UNEMPLOYED

of struggles so that analyses can be more widely disseminated, discussed, developed, and brought to have a more immediate bearing on strategy.

Grunwick: a '19th Century Sweatshop'?

. . . a sweat shop with a management which could have been lifted straight out of the Dickens era.
— Laurie Pavitt, Labour MP for Brent South, denouncing Grunwick in Parliament

Our members are being exploited by an employer more suited to the days of the British Empire than to the modern civilised state.
— Association of Professional, Executive, Clerical and Computer Staff (APEX) *Journal*, August-September 1976

Either this company moves into the 20th century, or it should go out of business. Either it agrees to negotiate or it goes out of business.
— Roy Grantham, General Secretary, APEX, April 1977

'Fort Grunwick', it was called, as day after day the television screens showed the strikers and their supporters 'besieging' the Grunwick photo processing plant in North West London at the height of the 2 year-long strike there. Eagerly recording the daily confrontation outside the factory, the media portrayed the strike support movement as victimising the 'loyal workers' who kept the factory running — the inside of which remained a largely private matter between the managing director George Ward and his employees. Thus the mass picket outside the factory gates could appear as a 'mob of howling hooligans baying for the workers' blood', while the police and the law could appear to be defending 'the right to work' by protecting 'the workers' from the picket.

Fort Grunwick, then, was not simply a factory but a base for an array of forces — police, judiciary, media — politicising an otherwise minor industrial dispute. As such, the strike was given epochal significance by those on both sides. For Keith Joseph, Tory Shadow Industry Secretary, it stood as 'a make or break point for British democracy'. For Hugh Scanlon, General Secretary of the engineers' union, it was 'a fundamental issue to the Trade Union movement, that we will fight to the very end to maintain'.

Roy Grantham, General Secretary of APEX, originally entered the strike with a rousing call to defeat this 'reactionary employer' George Ward, managing director of Grunwick, even by putting the firm out of business if necessary. Yet it wasn't long before Grantham was making concessions to Tory arguments about the management's 'right to manage' and the 'naturally' high rate of labour turnover in the photo-processing industry. Rather than Grunwick having to accommodate the union, in the end it was the union which attempted to accommodate Grunwick. Was the unions' shift, then, about trade unionism regressing back towards the 19th century, or about a new politics re-defining 20th century industrial relations?

APEX, whose motto is 'Militancy Within the Law', had originally cast its demands for union recognition at Grunwick within the employment protection legislation recently enacted as part of the Social Contract which the Trades Union Congress (TUC) had made with the Labour government. Although the issue of union recognition was to be formally resolved through state mediation by the Advisory, Conciliation and Arbitration Service (ACAS)[1], the mass action around Grunwick belied the unions' legalistic pretensions, at the same time that the unions lacked any consistent reply to the accusations that they were using illegitimate support actions.

The unions' dependence upon the state can be traced back to the beginning of their intervention into the strike. In August 1976 many of the Grunwick workers — mostly East African Asians, especially women — walked out over speed-up and compulsory overtime. Contesting management's control over them, the strikers quickly found their revolt narrowed into an APEX campaign for union recognition, ultimately through prolonged mediation by ACAS. The strikers faced a stark choice, with APEX giving them strike pay higher than their former wages, and Grunwick sacking them all and continuing the plant's operations without them.

In November, postal sorters threatened to cripple the firm by preventing mail order work from arriving there. However the Union of Post Office Workers (UPW) quickly ended the selective blacking when Grunwick apparently agreed to co-operate with ACAS' mediation attempts at the same time as the firm took High Court action against the blacking. Throughout the autumn and winter, the strikers and their (mostly Asian) supporters maintained an early morning picket which met harrassment from both the management and police, who made arrests to enforce their demand that the pickets limit their numbers at the gate to six only.

After months of disputation, ACAS finally recommended union recognition in March 1977, but Grunwick refused to comply. In June the strike intensified with the mass picketing organised by APEX with the Strike Committee and the local Brent Trades Council, led by Jack Dromey and Tom Durkin. Attracting hundreds of militant trade unionists, feminists and anti-racists as well as the strikers themselves, the mass picketing helped sustain the strikers' morale through publicity and fund-raising.

The mass picket was also intended to persuade Grunwick's remaining workforce to join the strike and to encourage other workers to extend the selective blacking against Grunwick which the Cricklewood postal sorters had initiated the previous November. More Grunwick workers did join the strike, and the postal sorters did briefly resume their selective blacking, but the mass picketing quickly became focused upon the scab bus which Grunwick introduced to get through the picket. The picketing took on a dynamic all its own, especially when attacked by a virtual police picket with brutalities by courtesy of the crack Special Patrol Group.

An aggressive police picket, daily television coverage and heated Parliamentary debates turned Grunwick into a national political issue. The stakes were heightened as the police moved to criminalise all picketing which fell on the wrong side of the line they drew between 'legitimate' and 'illegitimate' trade union activity. By the second week, the hundreds of arrests included Left Labour MP Audrey Wise and Yorkshire miners' leader Arthur Scargill, who had been bringing contingents of miners to the pickets.

The Tory Shadow Minister of Industry, Keith Joseph, denounced the mass picketing as 'blatant terrorism and illegality in the name of trade unionism'. The Labour Home Secretary Merlyn Rees, accepting the police's political initiative and defending their behaviour, complained that these disturbances had 'no place in responsible trade unionism'. APEX General Secretary Roy Grantham moved to shift the picketing back to the 'responsible' side of the line: he attempted unsuccessfully to make deals with the police on acceptable picketing, limit the numbers of pickets to 500 and impose armbands distinguishing official from unofficial pickets.

After four weeks of daily confrontations with the police, the Welsh and Yorkshire miners initiated a special mass picket for

July11. They aimed not simply to stop the scab bus but also to pressure union officials into extending the selective blacking already underway against Grunwick. This meant solidifying the postal sorters and cutting off essential services to the factory — a step deemed 'illegal' by Tory MP s, who thereby extended the threat of criminalisation from street picketing to traditional workplace solidarity.

In the event, on July 11 twelve thousand pickets defeated the police by blocking the factory entrances in the morning... only to divide over whether they should leave the factory gates in order to join the TUC's pre-announced march nearby — which most did, however reluctantly. It was 'responsible' trade unionism which won out by default, as the choice whether to remain or leave gave the pickets no collective means to confront the challenge of criminalisation.

Despite that innocuous result, the mass picketing and selective blacking now drew adamant opposition from APEX, accused of encouraging 'violence', 'illegal activity' and 'intimidation' (of Grunwick workers). When the government attempted to help the unions regain control by setting up an 'independent' investigation, APEX could argue that any further mass action should await the forthcoming report of Lord Scarman's Court of Inquiry. On August 25 Scarman recommended union recognition, only to see his report rejected by Grunwick. At the same time, Scarman's criticism of the mass picketing and selective blacking gave APEX a convenient way to beg off its earlier commitment to pursue direct action against Grunwick.

Appealing in vain to APEX for greater support, the strike committee did not decide to defy its own union publicly until October, when the strikers attempted to resume the mass picketing without APEX endorsement. By now the strike committee had lost political initiative to the narrow legalism of the TUC/Labour Party, which was able to deprive the committee of the full legitimacy which it needed to sustain its mass support. Subsequent mass pickets were even further criminalised by the police, whose Special Patrol Group brutally dispersed the November 7 mass picket — marking the effective demise of the strike.

Afterwards the strikers valiantly kept up a token picket for several more months. They also held a hunger strike outside the TUC headquarters to demand that the General Council make good on its pious promises of support — only to find their strike pay suspended. The strike was officially called off only the following summer, on 12 July 1978. The image of Grunwick as a throwback to 19th century anti-unionism persisted even with the strike's defeat. Tom Durkin, chairman of the Brent Trades Council, expressed a common view when he said, 'I am convinced that eventually George Ward will have to come into the 20th century and recognise workers' basic rights.'

What did this defeat really mean? This article will argue that the Social Contract, despite its aim of extending trade unionism, prepared the way for the unions' eventual deference to the apparently 'natural' demands of Grunwick's photo processing technology and markets, as well as for the political ascendancy of Grunwick as a veritable 20th century sweatshop. It is towards making such an argument that we will outline the consistent pro-Grunwick strategy developed by the National Association for Freedom (NAFF), as contrasted to the contradictory use which the unions made of the Social Contract, so that the potentially decisive strike support was politically marginalised from Grunwick's labour process. Then we will relate that labour process to Grunwick's chosen market and to its fixed capital, together well suited to exploiting its particular labour force. Then we will examine how the strike began and how the union recast the issues from exploitation to union recognition, leading on to the legal trap of ACAS and Scarman. Lastly we will explain how the support movement was politically defeated by forces which subsequently went on to establish a dominant consensus for a programme of extending the Grunwick model to British industry generally.

The NAFF Offensive: 'the right not to join a union'

> Power rests today not with Parliament but with the trade unions . . . The poor chap who is being exploited today is the man who's trying to run his own business.
> — George Ward, *Daily Telegraph*, 23 August 1977

> It is sometimes the employer, the job-creator, who is oppressed. Indeed, it is sometimes the non-union members who are oppressed — by the unions.
> — Keith Joseph, Shadow Minister of Industry

Grunwick's flouting of all the mediation procedures, dating from its initial refusal to co-operate with ACAS, owed much to

DEFEND WORKERS' RIGHT TO ORGANISE
THE ENEMY BEHIND
Gorst:
Sir Robert Thompson :
Brian Crozier :
Robert Moss :
N.A.F.F. the new employers association,
champions of sweat-shop
advising and representing
GRUNWICKS MANAGEMENT
DOWN WITH THE EMPLOYERS 'FREEDOM' TO OPPRESS
WE WANT A UNION
NO TO RACIAL DISCRIMINATION
GRUNWICK
Strike Committee
END
AVE LABOUR
£25 for 35 hrs
£28 for 40

the intervention by the National Association for Freedom. This organised force of the Thatcherite Tories provided a coherent strategic direction to the otherwise archaic anti-unionism of an isolated 'self-made' managing director, George Ward.

When the Cricklewood postal sorters first blacked Grunwick's mail in November 1976, Ward recognised their action as 'cutting the jugular vein of the company'. The NAFF's threat of a High Court injunction was accepted by the Union of Post Office Workers (UPW) as cause for ending the sorters' action. In due course the NAFF was to raise legal and political questions over the 'monopoly power' of the Post Office, the legitimacy of ACAS's mediation procedures, the boycott of chemist shops trading with Grunwick, the mass picketing yet to come and the proposed cutting off of water, gas and electricity services which never quite happened.

These challenges to the labour movement came at a time when the Social Contract was already dying as renewed wage demands eroded the Labour Government's credibility from within. The NAFF's campaign around Grunwick was posing a very different challenge to the Social Contract, from forces entirely outside the government — indeed, from outside the bipartisan consensus approach to industrial relations. For this reason the encouragement which Keith Joseph gave to Grunwick was countered by other Tories. Shadow Employment Secretary James Prior expressed the reservations of CBI elite capitalists when he urged Grunwick to comply with the ACAS recommendations on the grounds that the Conservative Party had long recognised ACAS as a neutral body.

Despite all its rhetoric of legality, the NAFF's intervention was less about simply deflecting political issues with legal technicalities than about using various legal procedures, police links and an all-too-willing media to *transform* the political issues into those favourable to the Thatcherites. The unions had acquired a special vulnerability by having chosen the law as their terrain. Now the NAFF could throw back at Ward's opponents their own accusation that Ward was flouting the 'rule of law'. The NAFF's intervention turned the unions' strategy on its head by representing the strikers' extra-Parliamentary pressure as somehow anti-Parliamentary and illegitimate, while itself using parts of the state (and the media) to sabotage industrial relations procedures and safeguards endorsed by Parliament.

The NAFF likewise turned the Left's 'Right to Work' slogan

on its head by attacking trade unions for denying individual freedoms to workers and employers alike.In particular their approach represented the strike supporters — even those inside Grunwick's labour process — as outsiders. A *Times* editorial in November 1977, referring to the blacking of Grunwick's mail-order work, helped to put the postal sorters on the defensive politically by echoing the NAFF's concern: 'What is so disturbing about the present case is that the postmen's interests are not directly involved at all . . .' The issues of Ward's freedom to exploit, and of his dependence upon the labour of postal, electricity and water workers, became overshadowed by doubts over the proper 'weapons of industrial warfare' (Ward's own phrase).

Furthermore, the NAFF formulated strategic initiatives for the anti-union forces, always keeping one step ahead of the support movement — indeed, structuring its very form, as with the scab bus. It was immediately after the highly publicised mass arrests on the first day of the mass picketing (13 June) that Grunwick introduced the scab bus, whose reckless driving through the pickets further provoked their anger.

The bus could not only get the 'loyal employees' into the factory but could also provide the media image of a 'howling mob' intimidating them. This tactic helped shift the political issues away from Ward's routine use of others' labour — both the Grunwick workers with their grievances inside the factory, and essential services workers with their actual or proposed selective blacking. Instead there was a spectacular conflict just outside the factory, with the bus inviting cries of 'scab' from a 'Rent-a-Mob' of unruly outsiders.

There was certainly a grain of truth in the otherwise 19th century image evoked of 'scabs' especially imported from outside to break a strike. With advice from NAFF, Ward developed a strategy of organizing politically inside the factory as well as outside, so as to create a 'loyal band of workers'. Ward at first hired the relatives of those still working, then geared up for the summer 1977 peak of the holiday snapshot market by recruiting workers from entirely outside the community of East African Asian refugees (e.g. Chinese students). As well as improving wages and terms of employment, he held 'loyalty' meetings every day to intimidate those who secretly wanted a union.

Not withstanding any superficial resemblance to 19th

century strike-breaking, these steps all formed part of the NAFF's modern political strategy of using the Grunwick strike to re-define 'the workers' as the loyal employees of a particular employer. By definition, this included only those workers willing to work on the terms which 'market forces' would seem to demand. Naturally the workers' future jobs depended upon obedience to the 'economic facts of life' in the highly competitive photo processing industry which 'by its very nature' required low wages and long hours.

The NAFF had chosen its field of battle wisely, since 'Fort Grunwick' was indeed well fortifiable. The factory enjoyed a police picket to guard it physically from the outside, and 'market forces' to impress upon the workers politically from the inside. The media eagerly took their cue from the NAFF in duly portraying the pro-strike and anti-strike forces as 'outsiders' versus the 'insiders'.

Thus Ward could credibly resist the bi-partisan consensus on industrial relations by appealing for his authority to the 'workers' inside the factory. ACAS, by asking the sacked strikers whether they wanted a union, was surely going 'outside of the company'. John Stacey, Grunwick's personnel manager, said of the sacked strikers: 'The truth is that we do not care what their opinions are, because they have nothing to offer the company.' (*The Times*, 8 December 1976)

Having helped to create this confusing situation, the NAFF knew well how to turn the strike to the advantage of its own political project. Although the attempt to re-define 'a worker' did involve legal technicalities, it did not depend on them, since the question it raised helped to disorient the strike supporters politically. The irresistible logic of the mass picketing lent itself to the NAFF's aim of representing the strike as a conflict between the Grunwick workers and the unions over 'the right to work', 'the right not to join a union', and 'intimidation' of the workers by the unions.

With the real intimidation coming from the Tory right, who were representing an industrial dispute as a political threat to democracy, there was more than just a grain of truth in the strikers' extending their cries of 'scab' from the remaining workforce to the police, the NAFF, and even Keith Joseph. Thus did they implicitly recognise that this was no simple trade union dispute, but did so in a way which attempted to subsume its strange political character back into the familiar trade union terms of reference. These terms had to be constantly re-imposed by a strike leadership appealing defensively to the presumed intrinsic legitimacy of an industrial dispute.

The NAFF, by contrast, holding the strategic initiative on all fronts, could afford to acknowledge quite freely that Fort Grunwick was less about a particular factory than about 'waging the class war by proxy'. Indeed the NAFF boasted (on a television interview) that this was really a 'dress rehearsal' for the next government. Yet the strike movement remained in fear of being seen to bring politics in to it, and so remained unable to formulate a political response to the NAFF's political offensive. By default the strike leadership began to accept the terms of reference which the NAFF had raised over peaceful vs. violent picketing, official vs. unofficial pickets — indeed, over insiders vs. outsiders. The postal, electricity and water workers became politically marginalised as the mass picketing itself became the central issue.

When the Cricklewood sorters took the mass picketing as an occasion to resume their selective blacking, the Post Office management locked them out for refusing to stop their 'discrimination' against Grunwick. Then, with the co-operation of the police and Post Office management, the NAFF's 'Operation Pony Express' quietly by-passed the sorters' blacking action by simply delivering Grunwick's blacked mail bags on 10 July 1977 — completely unopposed. The twenty thousand marchers at Grunwick next day allowed themselves no way to undo the NAFF's remedy or prevent its recurrence.

Coming from outside the government, the NAFF's intervention created a political framework in which sections of the state could take direct action entirely beyond the law — violating the statutory monopoly of the Post Office, and imposing a police picket — and so re-defining the law. By representing these illegal actions as the defence of the workers against the unions, the NAFF was able to throw the strike supporters on to the defensive as 'outsiders'. The NAFF's intervention brought the supporters to hesitate at being seen to act outside the 'rule of law' as re-defined by the NAFF. By political default the supporters' power *within* Grunwick's labour process was lost in the face of the NAFF's political clarity and organisational decisiveness.

The Social Contract: 'the right to join a trade union'

> We do not just stand behind you, we stand side by side with you all the way.
> — Len Murray, General Secretary of the TUC, speaking at a strike meeting in the Brent Trades Hall, 12 December 1977

Throughout the Grunwick strike the most potent weapon remained the selective blacking of Grunwick's postal, electricity, gas and water services. Yet the workers supplying those services — whom the NAFF kept politicially on the defensive — remained unable to recast the issue as Ward's dependence on their labour. Lacking a political basis for developing organic ties to each other or to the strike committee, they relied upon mediation by the Brent Trades Council, as the local organ of a TUC structured precisely to preclude such ties.

Manoeuvred into a defensive role of outside 'supporters', the essential services workers were left with scant organisational means to articulate politically the stillborn tactic of denying Ward the use of their labour. With the NAFF threatening to isolate each of them beyond the pale of the law, they demanded guarantees of support from the wider labour movement — as if such support could be committed in advance without their first acting to catalyze it. By demanding guarantees from the TUC, they invested in the TUC the power either to grant or deny wider support.

Solidarity was thus precluded by the need for guarantees which could not meaningfully be given. With each potential support group waiting for the others to take the decisive initiative, a chain of paralyses overtook the support movement and subordinated it to the union officials' authority. This chain extended from the essential services workers waiting for the strike committee to get a green light from the TUC, to the strike committee beholden to APEX for strike pay, to the TUC desperately avoiding any embarrassment to the Labour government, to the government offering the empty shell of the Social Contract as a way for the TUC to head off direct action.

The mass picketing outside the factory gates was supposed to be a way of breaking through the impasse. Yet, in itself incapable of making any decisive difference to the strike, the picketing became its central dynamic, posing the stakes for both sides more symbolically than materially. A picket which could never have been truly 'effective' generated a heated public debate over 'effective picketing'.

The mass picketing brought together all sectors of the strike's support in one time and place, but as 'delegates' to a symbolic event, without advancing the self-organisation and mass-democracy necessary to shut down Grunwick, either directly or indirectly. The picketing gave the strike supporters little alternative but to offer themselves as willing picket fodder for a theatre of the absurd on a stage not of their own making. Though spurred on by confrontations with the police, the mass picketing lacked the self-organisation and political clarity to become anything more than an appeal for the TUC and the Labour government to make something happen, which meant fulfilling the promise of the Social Contract. By demanding that the TUC and/or the Labour government take decisive steps to break Grunwick, the support movement in effect invested them with the power to do so or not to do so.

Under the Social Contract, which kept wage rises below the inflation rate, the working class had been suffering a decline in real wages and shopfloor power since 1975. Now all the frustrations of two or more years of industrial restraint exploded around Grunwick as the mass picketing and selective blacking undermined capital's command over labour. For the support movement, the slogan of 'the right to join a trade union' really meant the working class' right to organise, which held a contradictory historical content: inchoate impulses both to fulfill and abandon the promises of the Social Contract. While for the the TUC and Labour government, the slogan meant a conscious strategy of validating the mediation machinery as reliable guarantor of the much-heralded employment protection legislation, which was to have made the economic austerity tolerable.

In this context, the rallying cry to defend trade union rights suited the TUC and Labour Government's purposes by mystifying the historical nature of the threat posed by Grunwick. Ignoring the very modern character of both the NAFF's intervention and Grunwick's technology, the firm's critics attacked the management as a 'scandal'. Grunwick was simply an anachronistic throwback to 19th century sweatshops, a Dickensian-era employer stubbornly flouting the 20th century law of the land, in particular the Social Contract's employment protection law. So the task was simply to discipline a recalcitrant employer who was preventing the Grunwick's workers — and indirectly, other immigrant workers — from joining the Trade Union Movement and from enjoying the legal protection they were due. At the same time, Grunwick was seen as the thin end of the wedge, as signalling an anti-union offensive by the capitalist class as a whole, jeopardising the entire future of trade unionism.

However, the stakes were assumed to be about whether trade unions as such would be extended or regressed — whether history would move forwards or backwards — rather than whether the labour movement would depend upon legitimation from the Social Contract.Now that sections of capital were moving towards putting the nails in the Social Contract's coffin lid, the real historical stakes were about whether the labour movement would emerge weakened or strengthened from the impending death of the Social Contract. Its choice was either to resuscitate an increasingly lifeless Social Contract by demanding (however militantly) that the Labour Government make good on its promises, or decisively take the initiative away from the NAFF by building solidarity independently of the legitimation offered by the state and the unions. Unfortunately the timeless slogan of the 'right to join a trade union' obscured the concrete historical choices facing trade unionism. Thus the slogan helped trade union officials to identify the interests of the working class more closely with that of a government which claimed to be defending trade unionism against its Tory enemies.

Within the support movement the strategic debate was largely about what degree of militancy was needed to enforce the Employment Protection Act and so unionise Grunwick *through* the Social Contract. As compared to APEX's cautiousness, the greater militancy of the Brent Trades Council entailed a tactical difference over how best to pressure the government; the Council treated the mass pickets and selective blacking instrumentally, as a more effective way of ensuring enforcement of the Social Contract. In practice, this meant exceeding the legitimacy offered by the Social Contract, but without being able to develop the political understanding needed to supersede *dependence* upon that legitimacy in the face of NAFF's attempt to undermine it.

This more militant way of using the Social Contract ultimately came to mean being used *by* it — or more precisely, by the NAFF. By default even the support movement at times came to internalise doubts which the NAFF raised about the legitimacy of support actions, and so succumbed to the paralyses around Grunwick's labour process. Even those supporters whose labour was connected to Grunwick appeared more visibly in their role as 'supporters' located outside the factory, while the real outsiders — the NAFF — were intervening politically, though less visibly, inside as well as outside the factory.

With the stakes defined as whether an apparently 19th century sweatshop would be brought into the 20th century, the strike supporters tended to cast themselves in the role of outsiders demanding fair play for non-unionised workers. The defence they mounted against the NAFF's political offensive failed, despite its militancy, to articulate politically the inner connection linking the strikers to their supporters through the labour process.

Grunwick's Labour Process: 'quick, cheap & correct facts'

> Businesses like Grunwicks . . . could not have come into existence if they did not have a competitive edge, and the only edge they have is low labour costs and a willing labour force not restricted by trade union attitudes.
>
> — *Times* editorial, 30 June 1977

The Grunwick workers' grievances were in some ways typical of sweatshop conditions, featuring forms of 'absolute surplus value' such as compulsory overtime and speed-up. Such features would come as no surprise were it not for the popular image of technology as somehow inherently lightening the work load. In other ways, though, the grievances were bound up with the peculiar labour requirements of Grunwick's sort of product. By looking more closely at the grievances of the mail-room workers, we can see just how deep they run in the structural organisation of the firm.

In addition to the low wages (starting at £25 for a 35-hour week), there was also a compulsory overtime, written into the terms of employment, which was paid at only ¼ over the basic rate for the first six hours, and later only at time-and-a- half. Management could demand overtime (especially of the women) shortly before the normal end of the working day, without even permitting the workers to telephone home to arrange for their children to be picked up. By ordering the workers to record on job sheets the number of films booked in, the firm could also impose work quotas on the workers individually — a system possible only because the menial, routine character of the work lent itself to such quantification. The strict direct control over the workers went as far as requiring them, at one stage, to raise their hands to ask permission to go to the toilet.

The firm enforced its rules with the threat of sacking, quite freely used for relatively minor infringements, such as failing to report for work on a day of a bus strike. The management displayed a literally 'casual' attitude towards its work force in a way that would horrify more enlightened capitalists concerned with exploiting skills most effectively. For example, 'A woman is automatically sacked when having a baby and is told to reapply for work. This means that she has to start again at the minimum rate of pay.' (*Spare Rib* 54)

As *Financial Times* journalist Joe Rogaly put it in an eloquent understatement,

> The atmosphere suggested by the high turnover of staff in the mail-order department is one in which the development of a settled, long-serving labour force was apparently not a high priority. (p.38)

At Grunwick such practices need not be considered wasteful because the casual labour power is geared to a labour process that requires little of what might be called skill (or even continuity of personnel).

This is in turn no accident, but a consequence of the intentionally lop-sided way that Grunwick has come to be mechanized. The chemical processing and accounting phases have been subjected to an extreme automation that reduces its workers to machine-minders and reduces the clerical handling of film to a menial, highly labour-intensive routine. The computerisation of the accounts allows for rapidly retrieving information about money paid and owed by each customer, without clerks poring through file cabinets potentially under their human control. And the computerised chemical processing allows 360 films to be processed simultaneously without skilled technicians operating darkroom equipment. The computer can even reject defective negatives and so avoid printing them, and compensate for different light exposures within the area of a negative.

As a result, the clerical work of handling the film, especially mail-order film to be sorted and billed, becomes subordinated to the automated stages. Management can vary the numbers of clerical workers and their pace of work according to seasonal variations in custom , rather than automate the clerical work with a fixed capital that would remain under-utilised most of the year. The clerical work thus becomes the bottleneck for keeping up with the high turnover to which Grunwick is geared. And without directly subjecting that clerical work itself to the automation which indirectly structures it, high productivity requires a managerial supervision to maintain a continuously high pace of work in the mail-order department.

Now, while the time-pressure could be seen as due to the technology, it is more politically pointful to see both as derived from the class character of Grunwick's product. Grunwick's custom is not snapshot photos in general, but a particular section of the snap market which it helped to create. An unusually high proportion (70%) of Grunwick's custom is domestic mail-order, and that mostly amateur snapshots from working class holiday makers, with the remaining 30% coming from customers, overseas outlets and institutional work (including, appropriately, Scotland Yard).

Grunwick has successfully got the edge on its competitors from the sort of quick and cheap mail-order service it provides

our print n processing	usual price for processing 20-exposure film £	replacement Kodacolor film?	allowance for unprinted negative [1] p	cost for one reprint p	cost a print[2] p	time taken	overall quality of prints colours	sharpness
OUGH SHOPS								
ts[3]	3.45	no	14	19	22½	◩	□	□
ns	3.35[4]	no	14	18	22	◩	◪	◪
estons	3.35[4]	no	14	18	22			
k	3.90	no	16	22	24½	◩	◩	□
Smith	3.00	no	12	18	20	◩	◪	■
or	4.08	no	17	20	25½	■	□	◪
AIL								
uspool	3.05	yes	12	15	15½	◩	□	□
olor	3.45[5]	yes	13	15	17½			
ispool:						◪	□	□
lour Print Express	3.05[6]	yes	12	15	16			
e Film Service	3.05	yes	12	15	16			
erpost	3.45[5]	yes	13	15	18			
day Film Services:						[7]	[7]	[7]
scount Film Services	2.08	no	7	20	16			
itorial Film Services[8]	3.80	yes	16	20	19½			
mcare	3.50	yes	15	20	18			
ua:						◪	◪	◩
ail-a-Film, Postal Film Service, ght Price	2.40	no	7	14	17			
in-a-Film	3.35	yes	10	14	17			
mpost Laboratories	2.85[5]	no	10	14	19½			
:						◪	◪	◪
rclays	3.75[9]	no	9	—	19			
topost Express	1.95[10] [11]	no	5	16	15			
per Stamps	2.15[10] [12]	no	5	—	15½[12]			
sco	1.95	no	5	—	15			

so price charged for extra prints (eg 22 exposures 0-exposure film). Bonuspool and Holiday Film es (and Colour Print Express when you pay in ce) make no extra charge for this
cluding cost of a 20-exposure film and taking free postage and Green Shield stamps into account
ots use different laboratories in different areas of untry. Although they tell us they apply similar rds, our results are strictly valid for the North n area only
[4] Super Service: larger print size available: £3.80 for 20 exposures
[5] Payment when the prints arrive
[6] For payment in advance; for payment later, see Interpost charges
[7] Quality varies – see text
[8] Sold under the names of various magazines
[9] Processing paid for when film purchased
[10] Extra fast service available: 25p extra
[11] Larger print size available: 45p extra
[12] Including Green Shield stamps

KEY TO RATINGS ■ ◩ □ ◪ ■ best ⟷ worst

Which? magazine, June 1977

for colour processing — returning the film in just 4-5 days, as compared to 10 days to 3 weeks for other mail-order firms. The importance of this edge, as proudly noted by Ward in his book (p.24), can be seen from the way Grunwick's mail-order scheme (Bonuspool) is singled out for its fast service and low price in the consumer magazine *Which?* Of course, the class — in their fragmented role as 'consumers' of Grunwick's product — have little occasion for imagining what the fast service means for their fellow workers at Grunwick whom the management compel to work longer hours and faster according to the amount of work coming in.

As for why speed-up and overtime should be so integral to the firm, we must ask why the time factor for delivery should be so important. For an answer, we can look symptomatically at the product's use value, and beyond, to the tendency for 'instant' cameras to supplant the traditional mail-order market. With the 'use value' of the holiday snapshot already defined as a symbol of mass escape from the routine of jobs and home — both an advance and a palliative for the working class — it becomes important to recall the memories of that experience soon afterwards through instant nostalgia.[2]

While Polaroid and Kodak's instant cameras were doing the

use-value one better by making new capital investments entirely by-passing the mail-order procedure, newer small firms were investing in newly available colour processing technology. *Which?* magazine (June 1977) describes the advance,

> In the five years since we last tested colour film processors, the cost of living has more than doubled, but the cost of a colour print has gone up by only about a third. This has been made possible by the use of more and more mechanisation in the handling [processing] of films.

Joe Rogaly situates the opportunity seized by Grunwick:

> Grunwick was founded at a time when the work of printing and developing still films was leaving the corner chemists' shop and moving over to the large scale specialist companies. One of the reasons for this was the growth in the use of colour film. Small shops cannot afford the elaborate monitoring machinery which is necessary to process thousands of prints an hour, while yet providing something near to individual treatment for each one. A number of companies did well out of this technological change. (p. 31)

These technological and market advances had social determinants. The advances were possible only because Kodak's original advertising theme — the snap as 'cheap, quick and correct facts' — had now gone beyond its originally commercial market to dominate leisure as well, now subject to the same sort of time-economy as work itself.

But Grunwick actually began as a very different operation — processing only the black & white film itself (as many chemists had done as well), while sub-contracting out its colour work to Cooper & Pearson, with whom it later merged. According to Ward, Grunwick established itself, moving from a 'back street company onto a really viable basis' (p. 20), only by mass-distributing post-paid mail-order envelopes in newspapers and magazines, and by computerising the accountancy records from 'simple ledger posting machines'. (p.22) That change enabled Grunwick to launch a 'pay-later' scheme, 'Trucolour'.

Unfortunately for the firm, the management's premature expectations for the computer led to a crisis which required the invoicing of a record 200,000 mail-order colour films without the computer! (Ward, p.24) This meant having the workforce toil through the night to finish it — merely a more extreme case of heavy demands which Grunwick routinely made on its workers. The management made the necessity for sacrifice quite explicit;

workers took jobs there only as a last resort, leaving the firm as soon as they could find another job.

Though technologically Grunwick was by no means a 19th century sweatshop, its chosen niche in the capitalist relations of production and exchange allowed it little room for what *The Times* editorial called 'trade union attitudes' — collective bargaining over the terms of employment, and especially shop floor control over how the workers expend their labour power. That constraint set the task for management: how to enforce the sort of exploitation required.

Grunwick's Labour Force: 'It is not like Asian ladies . . .'

> Mr. George Ward, the head of Grunwicks, is a man worth a lot of respect . . . Mr Ward's company depends on a labour force with a high turnover and increasingly composed of immigrant Asians. A labour force like this can be either dumbly accepting or volatile. The machinery the company devised for handling any grievances if the workers proved volatile was sadly defective.
> —*Sunday Times* 23 August 1977

With Grunwick's chosen custom geared to a casual labour force, the managers were preoccupied with keeping them working hard. For example, Malcolm Alden — originally brought into Grunwick to set up the computer department — is later found in a glass-cased office overseeing the mail-order workers, whose work has nothing directly to do with computers, but ultimately everything to do with them.

As for questions of 'skill', which in the technical sense was hardly required for anyone employed at Grunwick, it is entirely possible that many of the mail-order room workers were more highly educated or trained than the managers, especially because that department's workers were drawn from local Asians who had recently been forced to emigrate from East Africa. In that region, where they'd been prohibited from owning agricultural land, they made at least an adequate living by working hard and long at running small shops (sometimes employing Africans) or by acquiring clerical and professional skills.[3]

There was an odd contradiction in their working for Ward in this country, since it was very much a role-reversal: they became the 'blacks', while their boss was an Anglo-Indian. Yet at the same time this work formed a part of each family's strategy for paying off mortgages on new homes or businesses here — in effect, a sacrifice towards their future economic independence.

The low wages were tolerated, at least initially, because it was usually the contribution by the wife, son or daughter to a family income which included debt repayments or university fees.[4] For example, when all the people still working at Grunwick during the strike were interviewed, it was found that only one worker was living solely on the income from Grunwick. (Rogaly, pp. 24-8) How convenient, then, for Grunwick to re-locate in Brent and get easy access to such an immigrant group, where the students will work hard during the summer and the regular employees are prepared to accept overtime whenever needed.

The control over labour power at Grunwick was rooted in sexual as well as racial oppression. Asian women are severely burdened with an ethos of 'female purity'. Their standing in the community becomes endangered by any unorthodox behaviour, which could also be seen as a threat to the 'male honour' of the husband or future husband of an arranged marriage. Amongst East African Asians (especially the Patel family, who had many members at Grunwick), there is a particularly strong pressure on daughters for fitness for marrying upwards — which entails a large dowry as well as a suitable public image for the daughter and future wife. Though threats of ostracism to wayward women come ultimately from the men, the norms of behaviour are enforced by women in the home. This burden weakens women's collective power, even when they are working together in the same factory.[5]

The few West Indian women employed at Grunwick are the exception which proves the rule, since they are kept out of the mail-order department and placed instead in the processing department. As machine minders, their (more moderate) pace of work is set by the automated film processing equipment, so that management need not risk provoking them with constant supervision. In *Race Today* (January 1978) Jayaben Desai suggests a social basis for the management's strategy:

> The West Indian ladies, who all work in the Processing Department, joined us late but are very active now. They are trying to find out everything that's been going on. They come to our meetings and give us ideas about what we should do next.
>
> You see, West Indian ladies — and I'm not blaming them — are less loyal to management than Asian ladies. Asian ladies are more loyal in the sense that they haven't got the same power in their hands as the West Indian ladies. Management knows this, or thinks in this way. In the Mail Department, money comes in every

> day in cheques and postal orders. One day in the summer season I collected £50 in cash and £20,000 in cheques and postal orders. Asian ladies are afraid to steal the money. Their conscience would not allow them to do that. They think if they catch us it would create a very bad impression in the whole community and they are scared of this. West Indian girls are not like them. They can do anything because they are stronger. So, probably, Ward has made this division and given the mail order jobs to Indian girls and process work to the West Indian girls on purpose. I don't *know* this is the reason, but it makes sense.

The Asian women's fears of ostracism posed a great obstacle for their revolt against Grunwick and for their full participation in the strike. One striker explained, 'It is considered a dishonour for a (Asian) woman to put herself in the public eye.' As an Anglo-Indian, Ward knew how to take full advantage of the Asian women's vulnerability in the factory and even on the picket line, where he threatened to spread rumours that the strikers were 'loose women'. Strikers described the management, and Ward in particular, as 'astute in their understanding of how to manipulate and humiliate them'. (*Spare Rib* 54)

So Ward 's hold over the workforce came not only from the workers' need for income but also — in the case of the Asian women — from a culturally sanctioned docility which could now be exploited by capital as the women began to work outside their families for the very first time. However, as became apparent, that docility had its limits once subjected to humiliation outside of traditional family entrepreneurships, especially in the political climate of 1976.

During the summer leading up to the Grunwick strike, a change of mood could be detected in the Asian community, especially in the face of racist attacks on the streets. For example a personnel officer at a Bradford mill was quoted as puzzled over the growing assertiveness of his formerly docile workers, '. . . it is not like Asian ladies to answer back or just walk away. I don't know what is the matter with them.' (*Spare Rib* 52) And the situation at Grunwick was about to strain the firm's system of control to the breaking point.

The Strike: The Human Bottleneck Cracks

> It had been the hottest, driest summer that London could remember... In the parks and on Hampstead Heath, people

> swam and took happy snapshots of one another as if they were by the seaside . . . Not far away, in Brent, business at Grunwick Processing Laboratories was booming. Most of their work is done in the summer, when the holiday film comes in to be developed and printed. It is like the ice-cream trade: sunny days create so much demand that those who work in it must move twice as fast as usual simply to keep in step . . . (Rogaly, p.9)

Rogaly's description of the breaking point is symptomatic of the way that the media, the judiciary and the unions tended to naturalize the social relations of production at Grunwick as technically determined. He takes for granted that the extra custom must take the form of compulsory overtime, and that leisure time will take the form of snapshot photos, and therefore speed-up. So his explanation must ultimately resort to determination in the last instance by meteorology. But the connection he draws between the weather and speed-up is well taken, and sets the context for the incidents which sparked off the strike in the mail-order department.

The first incident was a revolt against attempted speed-up by Devshi Bhudia, whose job was to sort geographically the packets of work before they were invoiced by the computer and prepared for mailing. Ward describes the increased work load in a revealing way, 'This was hardly a tall order. The reason for setting a deadline was at this stage of our peak season huge backlogs of work can build up fast unless we keep to a tight schedule.'

In this incident, workers rejected the attempted speed-up by overtly undoing their previous work, right before the eyes of the management. They continued working in 'slow motion', obviously sending up their supervisor but also commenting graphically upon the unusually intense heat (to which people in East Africa adjust by working more slowly). In this incident the speed demands of the market — itself partly created by Grunwick — presented an 'objective necessity' for a speed-up in the most labour-intensive parts of the process. That labour-intensiveness allowed, in turn, a resistance only by way of overt sabotage, since the workers' relation to the technology was an indirect one.

The second incident, on the very same day, was Jayaben Desai's refusal to accept compulsory overtime. Her job was to clear payments and complaints, and at 6pm she was ordered to stay on and continue checking lists of payments received from customers. In his book Ward explains matter-of-factly that 'it was always the company's policy to despatch all outstanding processed work on a Friday evening during the summer months before any of the staff left the premises.' (p.38) The consequence of computerising the accounts and mechanising the film processing is that the living labour in the mail-order department becomes the 'bottleneck' in the production process.

When Desai decided to walk out in protest, she announced to her fellow workers that she was resigning, not being sacked: 'They are taking advantage of us, they treat us like animals here. Just ask yourself why Mr Ward doesn't employ whites here. It's because they wouldn't stand for the pressure they [the management] put on us.' (*Race Today* April/May 1977)

In both incidents the management's demands may appear contingent upon extra-ordinary circumstances — which are, however, built into the class character of Grunwick's custom and technology. The firm and its product is founded on an especially racist and sexist exploitation of an immigrant workforce which is compelled to work longer and harder at routinely menial, measurable tasks. It was perhaps not for purely 'technical' reasons that Grunwick made it a priority to computerise the accounts and chemical processing, but not the handling of the film, where the living labour could be exploited all the more effectively.

With this perspective on the labour process, the structural weakness of the strike becomes all too apparent. Workers desperate enough to take such jobs in the first place would not so easily give them up, so the workforce was severely divided over the strike. And with a highly but selectively automated labour process, not organised on an assembly line basis, the plant could operate with a reduced workforce. This was especially true after the summer (1976) holiday peak, when Grunwick's custom normally declined anyway.

As for the support movement, the chemist boycott was highly successful in its own terms — reducing the number of retail outlets to only 1/5 its original total — but that boycott could only have a limited impact upon a firm with little trade normally coming in from retail shops. This feature made the postal sorters' temporary blacking of Grunwick all the more important, as a direct attack upon the time-economy (and therefore the use-value) of the firm's product.

The postal workers' industrial action was joined by other

2 THE SUN, Friday, July 1, 1977

Express is taken over in £13m deal

BEAVERBROOK Newspapers, which publishes the Daily Express, Evening Standard and Sunday Express, was yesterday taken over by Trafalgar House Investments.

The construction, shipping, property and hotels group paid £13,890,000 for the newspaper empire.

Mr Victor Matthews, Trafalgar's managing director, will become chairman and chief executive of Beaverbrook.

Mr Matthews said all three newspapers would continue to publish.

By and large, the editors will have complete freedom as long as they agree with the policy laid down," said Mr Matthews.

The proprietor, Sir Max Aitken, is to be invited to take the post of life president of Beaverbrook

POST REBELS GET NEW DEADLINE

POSTAL workers who refuse to handle Grunwick mail were last night given a new deadline to change their minds.

By MICHAEL GAY

The Post Office Board said they would put off action—which would include sending home the rebels without pay — until 12.30pm on Monday.

This followed a plea by Post Office Workers' Union leader Tom Jackson, who wanted time to "take new initiatives."

The Post Office also took note of Employment Secretary Albert Booth's call for a "calm atmosphere" to help the newly-appointed court of inquiry.

The rebels are 130 men at the post office in Cricklewood, North London.

They were originally told on Wednesday they had 24 hours to resume handling the blacked mail for the embattled Grunwick factory in Willesden.

And the Post Office Board warned last night that if normal working is not resumed by the new Monday deadline the original threat to suspend the men without pay will be carried out.

THE REBEL postmen will not be prosecuted, Attorney - General Sam Silkin ruled yesterday.

Such a move would be "wrong and counter-productive," he told the Commons.

There are people who would compromise with the devil

TORY MP RONALD BELL

Issue

The time had not come to take the issue out of the hands of the Post Office said Mr Silkin.

THE COMMONS emergency debate on Grunwick began with Tory lawyer Ronald Bell attacking Ministers for not condemning the rebel postal workers.

SILKIN ... prosecution moves "wrong."

Supporting Grunwick boss George Ward, Mr Bell said that for once the unions had found someone who would not crumble.

"There is always talk of moderation on this question. There has been too much of it, he said.

"There are people who would compromise with the devil. Gang warfare has taken over our country.

Eternity

"It challenges the rule of law and democracy."

MR WARD, who was in the Commons to hear Mr Silkin's speech, said later:

"If Sam Silkin waits until eternity he may establish law in eternity. But if I am dead what is the point? Dead men do not need laws to protect them."

THE SUN SAYS

Grunwick file their boo[ks]

By DAVID KEMP

GRUNWICK chiefs acted swiftly after the Government's warning that they risked prosecution if they failed to file their accounts.

Premier Jim Callaghan told the Commons yesterday that Mr George Ward had now sent them in.

It was only on Wednesday night that Junior Trade Minister Stanley Clinton Davis issued the warning.

He said the firm had not filed accounts for the last two financial years, as required by law.

Law

Yesterday Labour left-winger Mr Dennis Skinner, urged the Prime Minister "to take time out to examine the company accounts of that squalid little man Mr George Ward, who apparently, has not sent accounts to Comp[anies] House in comp[liance] with the law."

Mr Skinner ask[ed the] Premier: "Do yo[u] get a little sl[...] watching George [...] on television, with some of hi[s col]leagues on the benches, pontif[icating] about upholding letter of the law [...] he is not even ca[rrying] it out himself"

Mr Callaghan r[eplied:] "I understand h[e has] now complied by [sub]mitting his accoun[ts]

FRONT IN A ROYAL ROW

AN invitation to two members of the National Front to attend a Jubilee reception for the Queen has sparked a storm of protest.

The Front members will stand in a special VIP enclosure with other local politicians when the Queen goes walkabout in Islington, North London, on Wednesday.

Their invitation came from Islington's Labour-controlled council.

It was attacked by local Labour MP John Grant, who is Minister of State for Employment.

Appalling

He said: "This is absolutely appalling. I see the National Front as a neo-Nazi group dedicated to stirring up strife in what is a multi-racial community.

"I have sent a message to the mayor asking him to reconsider the invitations because they are causing a great deal of ill-feeling."

Mr Grant said he hoped the Front members — prospective Parliamentary candidate Paul Kavanaugh and Mrs Thelma Berry — would not now accept the invitations.

"It would cause considerable embarrassment," he said. "The best thing they can do is to return the invitations."

WEATH[ER]

MOSTLY cloudy, rain.

OUTLOOK: Cha[nge]. Becoming warmer

London, SE, Cen[tral] E Anglia, Midl[ands] dry. Max temp [...]

SUN rises 4.46 [am, sets] 9.21 pm. LIGHT[ING-UP] pm to 4.18 am. WATER London [...] 7.40 mg.

COOL OUT[LOOK]

July will be [...] changeable, the [...] predicted yeste[rday ...] is expected to b[e ...] than usual.

THERE ARE MORE CLASSIFIED JOBS ON PAGE 23

On 1 July 1977 the *Sun* appeared with a blank space instead of its Grunwick editorial after printworkers raised objections. A few days earlier, the June 25 edition of the Sunday *Observer* had lost ¼ million copies through delays when the printworkers protested at an ad placed by the NAFF entitled 'Union Private Armies Have No Place Here'. The introduction of new technologies currently being resisted by Fleet Street print workers will make it more difficult for them to intervene in these ways.

workers who made similarly selective withdrawals of their labour:[6] airport and dock workers who blacked Grunwick's continental mail-order work; National Union of Bank Employees members who attempted to black Grunwick's accounts; TGWU members who refused to drive police coaches to the mass pickets; college teachers who successfully protested at the police use of facilities at Willesden Technical College (conveniently adjoining the Grunwick factory). All this was in addition to the thousands of workers , especially miners, who left their jobs for a day (or more) to attend the mass picket. In London, the Camden District Works Department workers went as far as to call a strike for June 17 in support of the Grunwick strikers.

Many have already pointed out that these actions, taken in the name of 'supporting the right to join a trade union', were often taken in a way oblivious to the specific racial and sexual conflict that gave rise to the strike. As one Yorkshire miner remarked in a sequence included in the radical film *Stand Together*, the miners had come to support the 'lads' at Grunwick. This remark was cited in an ironic vein in a feminist review of the film. (*Spare Rib* 66)

However it has not been pointed out that those racial and sexual aspects of their class oppression were so integral to the labour process that no legal machinery — however sympathetic to the workers — could do 'justice' to it. Let us examine more closely APEX's complicity in the way that the legal procedures abstracted out 'economic' issues from the relations of production.

APEX: 'Militancy Within the Law'

> Britain must aim at an equal opportunity economy in the 1980s. Much else will be needed too as a basis for national revival — high efficiency, low costs, bold investment, good industrial relations. But an important ingredient must be equality of opportunity. This is not just a desirable objective on grounds of justice and morality; it is in the obvious self-interest of every organisation to make the most of its available talent and to ensure a contented workforce.
> — David Lane, Chairman of the Commission for Racial Equality (CRE) speaking at the TUC conference on Race Relations, July 1979, as reported (uncritically) in the printworkers' *SOGAT Journal*

> This is the first major dispute involving immigrant workers to which the unions have given *full support*.
> —Roy Grantham, General Secretary, APEX, March 1977

In strikes preceding the one at Grunwick, Asian strikers in Britain had been building up community support, especially fund-raising. Those strikes directed against racist hierarchies at their workplaces were often refused support by the trade unions, which defended the divisive privileges of white workers. As Asian strikers drew instead upon a community support tradition from the Indian sub-continent, the trade unions responded by reprimanding them for failing to follow the 'proper dispute procedures'.

By 1976, however, the unions were taking fright at the strikers' community base as a threat to the officials' authority. A local APEX official, Len Gristey, expressed such fears about the Grunwick strike when he reported to Roy Grantham during the first week: 'A further difficulty is that militants among the Asian community are seeking to get hold of the matter for their own political purposes and this has to be watched very carefully indeed.'

It was to establish the credibility and authority of the 'proper procedures' that trade union officials made their dramatic gestures of support. APEX offered strike pay exceeding the Grunwick's weekly wages; the APEX General Secretary made Grunwick a major item at the TUC Conference only days after the dispute began; and the TUC General Secretary made an unprecedented appearance at a local support meeting in December 1976. But those tactics were devised not simply to sustain the strike but to take it in a particular strategic direction — subsuming the workers' grievances under the TUC's Social Contract.

If we look at how APEX first became involved as it did in the Grunwick dispute, we can see that the union looked to the dispute for extending the legal framework of the social contract to the realm of racial oppression, treated as a special legal problem. At Grunwick this became necessary because, as the critique in *Race and Class* put it, 'the dispute had called into question the protective employment laws . . . which had been the trade-off with the government in return for the Social Contract'.[7] At Grunwick the unions were determined to make those laws work. 'To bring the dispute within the Social Contract framework, it was first necessary to unionise the Asian strikers. But to unionise a black workforce, it is first necessary to take a stand against racial discrimination.'[8]

When Roy Grantham of APEX spoke about the Grunwick strike at the September 1976 TUC conference to solicit support, he did so by way of speaking to the Government's White Paper on Racial Discrimination: 'Here is a clear case of a reactionary employer taking advantage of race and employing workers on disgraceful terms and conditions.' The suggestion that these conditions are separable from the production process underlies his third demand: 'that all those who have been sacked should be reinstated, and that the union should then proceed to negotiate for an end to the deplorable conditions'. (*TUC Report* 1976)

In the context of the Social Contract, APEX's intended role was to treat racism at Grunwick (and elsewhere) as a legal problem of 'racial discrimination', as an extra-ordinary injustice inflicted upon individuals — its location in a factory being largely incidental. APEX intervened so as to reduce this politically explosive aspect of class struggle to terms by which it could be quantified and negotiated around a conference table. In other words, as *Race and Class* put it, the state could hand over to the trade unions 'the management of racism in employment', as part of 'selling the Employment Protection Act to the workforce as a whole.'[9]

Of course, the industrial action taken later by other sections of the working class threatened to break out of the legal framework that APEX had intended to legitimate. But after it became clear that the extra-legal support movement had been contained by the unions and the state, one striker remarked, 'Even if APEX is recognised, it will not be *our* victory.'[10] For at that stage union recognition would not promote the shopfloor revolt that had sparked the strike, but would help fulfill Roy Grantham's aim of extending to racial discrimination the existing legal procedures and statutory safeguards of workers *as individual citizens*.[11]

To see the formative stages of the ultimately abortive Grunwick branch of APEX, we can look at the way the Scarman hearings cast the union as legal advocate rather than as collective power. For example, here is Lord Scarman cross-examining Jayaben Desai (treasurer of the strike committee) on 12 July 1977 on her reasons for wanting a union:

> What you are saying is that if you have someone to speak for you and someone who is on your side who is independent then you have a greater sense of security?
>
> Yes, and that is what we want.

> It might interest you to know, Mrs Desai, that is the origin of the legal profession.

Of course, Ward himself claimed to hold no objection to this innocuous role for a union, as he stated in his reply of 31 August 1977 to the Scarman Report, '. . . the Company . . . has always been prepared to allow any employee who wishes it, to have a trade union bargain on his or her behalf.'

In this respect it is significant that Joe Rogaly (of the *Financial Times*) described the strike as somehow not at all fundamentally about class struggle, which was brought in from the outside, starting with the left-wing Brent Trades Council. For example, Rogaly speaks of the 'really important events, the ones that began to bring the British trade union movement and the British class struggle into this picayune revolt . . .' (p. 18) In fact, he ascribes the volatile behaviour of the strikers not to the humiliating working conditions but to their unfamiliarity with trade unionism! 'But they can be forgiven for this lapse from normal trade union practice on the grounds that they could not be expected to know better, being recent immigrants.' (p. 163)

Such a patronising attitude finds resonance in the Scarman Report itself, which considers the strikers 'ill-acquainted with the conduct of industrial relations in Britain.' Even Jack Dromey of the Brent Trades Council said as much when he testified at the Court of Inquiry that the strikers at the outset were like a 'bunch of chickens lost outside of a coop'.

The Legal Trap: the 'Value of Labour'

> If we took these dismissals to an industrial tribunal we would be confident of winning reinstatement. This is a clear case of unfair dismissal as defined under the Trade Union and Industrial Relations Act 1974.
> —Jack Dromey, Secretary, Brent Trades Council, September 1976

> Why not use both Section 11 of the EPA which lays down a clear procedure for union recognition and the authority of ACAS?
> —Jack Dromey and Tom Durkin, Brent Trades Council, February 1977[12]

When the Labour government moved to end the direct action against Grunwick by imposing an 'independent' inquiry under Lord Justice Scarman, the Brent Trades Council and the strike committee opposed APEX's reliance upon such an inquiry rather than direct action, whose original rationale had been the months of ineffective legal procedures. However, once the unions had intimidated the strike committee into suspending the mass pickets (and the postal sorters their blacking), there was a tendency to look to the 'legal road' yet again for vindication. Dromey himself left the way open for staking the legitimacy of further militant action upon the Inquiry's verdict when he warned that 'The Inquiry will not resolve the dispute unless the pressure from the trade union movement on the company is maintained and escalated.' (Dromey, p.155)

Though there was the obvious danger that the timing of the Inquiry would make it into a substitute for that pressure, there was a less obvious danger that the process of winning vindication within the Inquiry would draw the strikers into accepting the 'market realities' which had made their exploitation 'necessary'.

The Inquiry represented the Grunwick dispute as merely about racial discrimination, not class struggle, and so remediable by legal procedures that would rectify abuses in an otherwise technical process of production. The validity of the workers' grievances was judged by whether the protested conditions were seen to be arbitrary in relation to a base line set by other such firms. Was there racial *discrimination* against Grunwick workers, collectively or individually? This is what it meant for the Scarman inquiry to ask whether, as Grantham had charged, Ward was 'taking advantage of race'. And this is what it meant to try to win merely a union, however militantly, through the law — which had to take as a given the technology and concomitant relations of production of the photo processing industry.

Rogaly describes the dilemma of legal experts making statistical comparisons among similar firms — an exercise that mystifies the social conflict lived by the workers every day: '... the difficulty is that direct comparisons between one company and another are always awkward. Working conditions, overtime, subsidized meals, and scores of other factors come into it. "I am advised we never will find exact comparables", sighed Lord Justice Scarman on the ninth day.' (Rogaly, p.33)

In the end, the clean, modern, *normal* Grunwick factory passed Scarman's technical inspection. The law could understand the social relations of production only as quantifiable parameters, certainly not as a social control of labour for production. Relations take the form of measurable things:

'Physical working conditions were reasonably good', says the Scarman Report, as if these 'conditions' were some external context or environment *within which* production occurs, and not the material process of production itself.

Underlying this line of inquiry was the notion of 'exploitation' defined in purely relative terms as compared to financial and technical necessity in the industry as a whole (as in the pro-Ward editiorial in *The Times* of 30 June 1977). Were Grunwick's wages and working conditions objectively necessary, or arbitrary? Such terms were echoed in John Andrews' *Guardian* article of 23 June 1977, the only analysis published that considers the nature of Grunwick's custom: amateur holiday snapshots. His article asks whether Grunwick's wages were low compared to other such firms or even compared to the (low) price of the firm's product. Of course the question assumes that the workers are selling their *labour* (its product technically fixed) and then asks whether — as individuals — they are unjustly denied the full value of that labour.

Neither the capitalist press, nor the Scarman Inquiry, nor the trade unions, could ask how racism enters into the management's methods of compelling the workers to expend the *labour power* that they are selling. While considering the workforce as a whole, the Inquiry's mode of investigation aimed to treat the social relations of production at Grunwick as an aggregate of individual labours possibly receiving unfair treatment and compensation, rather than as a power relation over production.

The Scarman Inquiry's task at hand was in effect to restrict the issue of racism in the plant to one of the 'economic' — the unfair remuneration and 'physical' working conditions. The Inquiry process excluded from the question of racism any consideration of the control of labour power in production. In this way the capitalist social relations of production could be contained within the separate forms of economic and political — 'economic' grievances subject to statistical analysis and mediation by a third ('political') agency, perhaps invoking a trade union as its own creature, with 'political' pressure being applied to the state to recommend one way or the other. The control of labour power was to remain a given, technically determined by the forces of production, with only the 'value of labour' in dispute.[13]

So, when the Scarman Report speaks of the law as 'a curb

upon power — irrespective of the person or institution who wields it', or when Rogaly hails the law as an alternative to 'rule by those with the greatest muscle' (p. 162), a distinction between different forms of power is being lost. The Scarman Inquiry asked questions which *presumed* capital's power to regulate labour power according to its exchange (market) value. The Inquiry helped to defuse the power of the working class to refuse such regulation of its labour power, at Grunwick and elsewhere. But the limit of such state intervention appeared when the employer Ward refused to comply with what could only be recommendations. It is precisely because the law is built upon exchange relations that it cannot set out to 'enforce' them.

Just as the legal apparatus could not consider the nature of Grunwick's product and the methods of producing it, neither could it force Ward to recognise APEX, but could only ask him to do so in the general interests of his class.[14] As the Scarman report had to admit, Grunwick 'has acted within the letter but outside the spirit of the law'. The Court could only ask that Ward comply with the spirit of the legal machinery for the sake of 'good industrial relations'. It pleaded, 'If men act unreasonably, by which we mean in obedience to the letter but not the spirit of the law, it will not work'.

Ward noted the apparent inconsistency between the Court's assessment of Grunwick's working conditions and the recommendations against the company. 'The evidence, as expected, favours Grunwick. The recommendations, as expected, do not.' Ward claimed (rightly) that the Court of Inquiry 'had been established for a political purpose . . . Its job was to resolve a dispute which was embarassing to the Government.'

However, against what Ward saw as an 'encroaching corporate state', he counterposed 'natural rights' and 'individual freedoms'. Apparently he failed to realise that the capitalists' freedom to exploit rests tenuously upon a foundation of legal concessions granted to the power of the working class, which (he resents being told) 'it is most unwise to resist too strenuously'. When the Court of Inquiry found, inevitably, that Grunwick's statistical parameters did not differ markedly from other such firms, yet still recommended union recognition, Ward would not accept the adverse ruling as necessary for stable class rule.

While Ward rejected the Report, it is important for us not to accept the *terms* of the report. Through the legal procedures of

ACAS and Scarman, APEX and the Labour Government left racism essentially unchallenged. They abstracted it from production by casting the workers' grievances in the legalistic, technically definable terms suitable for court inquiries and union negotiations, suitable for disputing 'the value of labour' and its free and fair exchange.

The fragmented and routinised division of labour at Grunwick, facilitating the management's manipulation and calculation of the labour power it buys, is no exception but a universal tendency of capital towards reducing all labour to interchangeable, abstract labour power. Bourgeois law can do no more than to defend freedom and equality *within* that historical tendency. Accordingly, the Grunwick workers' exploitation lent itself to legal procedures only at the expense of narrowing the issue to sterile technical disputes about 'exact comparables' with which to identify possible racial discrimination. Unlike the real revolt underlying the Scarman Inquiry, such quantification cannot challenge — but must reproduce — Grunwick's relations of production.

The Support Movement

> If we can't crack Grunwick, we can't crack anything.
> — Tom Jackson, General Secretary, Union of Post Office Workers (UPW), 13 June 1977

> If we lose this one, we will lose them all.
> — Chris Wright, Westminster branch secretary of APEX

If the legal procedures remained just barely credible enough to defuse the mass support actions, it was because the threat Grunwick posed to the entire working class was seen as a sequential one, according to a 'domino theory' in the area of small enterprises, rather than as a broader threat immanent in the NAFF's political campaign against 'trade union attitudes'. The Scarman Inquiry's supreme success lay most fundamentally in abstracting the Grunwick strikers' grievances out of the labour process that perpetuated them, thereby severing the inner connection between their labours and the essential service workers', likewise between their respective revolts against capital's command over labour. By endorsing trade unionism, Scarman could credibly demarcate a boundary between 'legitimate' trade union disputes over the value of labour, on the one hand, and the 'British disease' of shopfloor power and class solidarity, on the other. One result, for example, was to reinforce *The Times*' claim that 'the postmen's interests are not directly involved at all'.

What was really at stake was not simply the extension or regression of unionism as such, but the political direction it would take with the disintegration of the consensus 'spirit of the law' on which trade unionism had become increasingly dependent since 1975. However unfortunate for the state corporate strategy, Ward rejected the Scarman recommendations. But this was something more than a personal anti-union prejuduce, if something less than a class-wide anti-union crusade; this was the necessary position of a particular form of capital (and production relations) embodied in the Grunwick plant and personified in George Ward. The workers' grievances, especially the racism, was so integrally institutionalised in the plant's labour process, that not even Scarman's recasting of them could bring the management to accept union recognition.

Yet to demand more resolute action against Ward (especially nationalization) was to subordinate the power of the class to a state-granted unionism adapting to Grunwick's relations of production. Furthermore, it was to deny the supporters' own class power by appealing ultimately for their legitimacy to the Labour government, which needed first and foremost to fragment the struggle into a narrowly economic trade unionism and a narrowly legalistic politics — even if both contained simultaneously within the Scarman Inquiry. Generally missing among the support movement was a sense of how the terms, terrain and stakes of the struggle had been set by the NAFF's historic opportunity with the decline of the Social Contract to which the trade unions were appealing in defense.

With the NAFF's campaign around Grunwick, the original project of winning trade union recognition *through* the Social Contract became a historical impossibility. There could be no middle ground of pressuring the state, however militantly, because the support movement's militancy was further undermining the Social Contract from within, weakening the Labour Government's claim against the Tories that it could best contain the unions. The state's priority necessarily became one of reasserting its own authority over 'public order', while defending the mediation procedures in the name of defending trade unionism. The unions, in turn, having staked their future upon that of the Labour Government, obligingly helped to contain the support movement by policing the boundary between economics and politics within it; in particular the UPW disciplined the Cricklewood sorters for taking at face value their union's rhetorical promise to support the Grunwick strikers.

While the unions recoiled in horror when the support movement began to jeopardise their deal with the Labour government, the left was unable to speak to the power of that movement as anything more than a pressure tactic for winning trade union recognition.[15] Defending 'the right to join a trade union' from right-wing attack could not make historically self-conscious the revolt already underway against industrial restraint both inside and outside the Grunwick factory; it could not counterpose any revolutionary aspirations to those expressed by the TUC for trade unionism to be legitimated by the state.

By posing the stakes in much the same *defensive way* as did the unions, the left gave itself little means of challenging the support movement's ultimate 'unity' in submission to the TUC's restraint. It could not even mount an *authentic* defense of those who had taken the greatest risks.[16] The rhetoric of 'defence' served to ensure that the unions would retain ultimate power to 'betray' — or rather, to express social-democratic aspirations, organisationally unopposed.[17]

The Twentieth Century Sweatshop: 'a prosperous and efficient company'

> We're not seeking to interfere with his [Ward's] Rights at all; all we want to see is the reinstatement of our members and good relations established with this company because that would be in the interests of our members, whose interests are the important thing in this issue, that their standards be improved. And in order to do that, we must have a prosperous and effective company. And we will work with them to that end . . .
>
> The reality is, he's got a good company, and he needs our members, who are good workers for him.
>
> We will say to him, look, as natural turnover takes place in your company, and there's a high rate of turnover naturally in this industry, we want our members to fill those vacancies, as you expand the company once again. We want our members to come back, good people, some of whom have been told during the course of the dispute, "if you'd stayed with us you'd be members

of management today" — that's how much they appreciate some of our members.
—Roy Grantham, General Secretary, APEX, interviewed by Brian Walden on *Weekend World*, 16 October 1977

The disastrous strategy of 'defense' against Grunwick was grounded in a broader treatment of Grunwick as a marginal, atavistic anomaly. Certainly, with its immigrant labour force employed in a 'leisure' consumption industry, Grunwick may well seem to stand outside the mainstream of modern capitalism in Britain. And such an impression has been perpetuated by the misplaced references to 'another Grunwick' each time immigrant workers have struck against anti-union employers (e.g., Garner's Steak House, Trust House Forte Hotels, Chix confectioners, etc.). This is as if the Grunwick episode were about a marginal workforce trying to 'catch up' with modern industrial relations.

However, an apparently marginal firm like Grunwick appears integral to modern capitalism once we locate it historically within the forces and relations of production and reproduction. Grunwick exists as a consumer of advanced computer technology, as a producer of commoditized working class leisure, and as employer of an 'internal colony' with counterparts all over the world, from Silicon Valley in California to the Free Trade Zones of Southeast Asia.[18] There can be little effective opposition to the 20th century sweatshop short of confronting the technology itself, the use values and consumer markets to which it is geared, and the international division of labour which it structures.

At Grunwick the truly docile 'dead labour' of modern science and technology joins with work discipline and management attitudes which recall the earliest struggles of the labour movement. Yet all of this is now, in the late 20th century, in the service of a 'society of the spectacle',[19] where commodity consumption and re-lived fantasy consumption are consolations that capital offers to palliate the labour force. As Paul Simon says in his song, 'Everything looks worse in black and white', while Kodachrome gives you nice bright colours — 'makes you think all the world's a sunny day.'[20]

At the same time , Grunwick's modern character made it an ideal choice for the NAFF's project of prefiguring a new Tory government of May 1979, whose anti-union offensive extended the Grunwick test case to a new industrial strategy, that is:

—from one employer's defiance of the Social Contract, to the near-collapse of the bi-partisan consensus on industrial relations, particularly in the debate over the Employment Bill;
—from 'Rent-a-Mob', to restrictions on 'secondary' picketing, a newly coined phrase to demarcate 'insiders/outsiders' within the same labour process;
—from the police-imposed limit on pickets, to the statutory re-definition of a mass picket as a 'demonstration' (and therefore outside of industrial relations law);
—from legal action against postal sorters, to proposals to end the 'monopoly power' of the Post Office, and hence of postal workers;
—from 'the right to work', to the attack on shop floor power in the name of protecting workers from 'union bosses' (e.g. by sacking Derek Robinson, shop stewards convenor at British Leyland);
—from the competitive necessity of the 20th century sweatshop at Grunwick, to the necessary sacrifices of restructuring mainstream British industries for international competition (e.g. British Leyland, British Steel);
—from Grunwick's 'loyal band of workers', to British Leyland's appeal (over the heads of the union) for the carworkers to produce cars as the market demands, as with the Mini-Metro.

The Grunwick struggle has been popularly celebrated as a shining example of highly organized, highly disciplined, highly paid white workers coming to the aid of underprivileged workers, in the sense of uplifting them into the trade union movement. This article has tried to show that, despite all the heroic militancy, the dominant historical movement ran in just the opposite direction: an anti-working class offensive extending from more marginal to more mainstream sections of the working class, with, for example, the UPW effectively accepting capital's definition of 'secondary' blacking. Grunwick became the unions' *cause celebre* to which the strikers, and then the supporters, were sacrificed.

It is only such defensive self-policing by the unions which made it possible for capital to envisage enforcing a new 'Employment Act'. Indeed, under the Tory government, even the existing law worked increasingly against the class-wide solidarity of 'secondary' picketing and blacking as workers — by virtue of their own hesitancy — invested more power in the existing law to decide what they could and could not do.[21]

Capital's project here is to decompose the working class as a class, politically fragmenting its collective relations of production into purely market relations among different workforces, each subordinating its internal production relations to market competition.

For example, British Leyland's successful sacking of Derek Robinson — truly 'another Grunwick' — aimed at realizing the full disciplinary potential of the assembly line in automobile production. It was in the name of market competition that BL Chairman Michael Edwardes set out to re-asssert management control over line-speeds, extending automation to a degree incompatible with 'trade union attitudes', or the resistance aspect of the shop stewards' mediating role. Here the BL management took a calculated risk in gambling on winning more control by repressing the unofficial movement than by incorporating it.

At British Leyland and Grunwick, the official unions played the official role of adjudicating the legitimacy of the unofficial movement, which largely deferred to the unions' prerogative. As capitalist opponents of the Social Contract's incorporation strategy moved to withdraw legitimacy from the unions, the unions responded by attempting to prove themselves worthy of it, by demonstrating their capacity to police the production relations of the 20th century sweatshop.[22]

In this sense the Social Contract was ultimately used to greatest advantage not by the working class, but, indirectly, by the Tories, who won the power to withhold or restore legitimation to the unions. At Grunwick, if there was any regression to the past, it was not the firm's management but the strike support movement, looking to a disintegrating social-democratic apparatus. Grunwick, far from a vestige of the 19th century, turned out to be capital's vanguard of the late 20th century, politically and technologically.

Indeed, capital is taking the political initiative in posing the question of what we produce, how and why. What is labour's response? As mainstream industries succumb to the Grunwick model, how do we combat the 20th century sweatshop writ large?

It was in March 1979 that the 1980 Corporate Plan was agreed. Two months after the Tories came to power, the original 1980 plan was scrapped and new proposals put forward. The Confederation Executive were bullied into accepting the proposals, as in their famous statement about being taken to the precipice.

It has become clear in the past few months that the real purpose of the ballot, that the real purpose of the publicity campaign was not to push through a plan to save BL, but was to foist upon the workforce a document which I believe Sir Michael has used and will use to break the trade union movement in BL. The document is this: it's called the Bastards' Charter, because it would establish within British Leyland, and subsequently in every motor car company and every engineering company in this country, a system of management that will put us back 100 years. Read this document, because it sets out the destruction of your organization.

— speaker at rally held after Derek Robinson's sacking, November 1979, Birmingham Town Hall

Now if we were to deliver a plan to the government where the workforce *co-operate totally*, where we feel we can put forward a viable plan, it would still remain to convince the government that they should back that plan. So I want to say here and now: this isn't a deal. I'm saying frankly through you right now to the workforce: for goodness sake *co-operate fully* in this programme. It's our only hope.

— Michael Edwardes, Chairman of British Leyland, speaking on television.

THE TIMES TUESDAY NOVEMBER 11 1980

Why the unions are keeping their heads down

Paul Routledge examines the new mood of union moderation

Brought up against an intransigent government and the frightening reality of mass unemployment, it seems that the bubble of union power has been well and truly pricked. After the Robinson affair, shop stewards are keeping their heads below the parapet. In today's bargaining climate of fear, single-figure wage settlements are commonplace, and management such as Fords feel confident enough to bring back stronger shop floor discipline. The CBI is abandoning its joint commitment with the TUC to agreed guidelines for the introduction of new technology. Employers will simply install sophisticated labour-saving systems, it must be presumed, whether the unions like it or not.

Postscript

Readers may feel that this article sits strangely in a collection on science, technology and the labour process, but the strangeness is not merely a basis for criticising the article. The extent to which the body of its exposition takes up politics in a more traditionally recognised sense can serve as an implicit challenge to the other articles in the collection. It invites the authors of studies on science and technology to pay a great deal more attention to how their analyses of particular knowledges and technologies articulate with the social relations of organised class struggle.

The other sense in which this article sits strangely in the collection is complementary to the above point. That is, the article may have over-emphasised the organisation of class struggle at the expense of the fundamental point. It is this: high technology, and the latest fads in consumerism, do not make the sweatshop a Dickensian anachronism; rather, they give it a new lease on life.

In particular, capital will retain labour-intensive tasks, even within a relatively capital-intensive production process, whenever such an arrangement seems to offer greater profitability. Likewise, capital will replace living labour with dead labour (technology) only insofar as the greater investment costs are outweighed by the greater advantages of exploiting the living labour which remains. Though this trade-off does operate in peripheral industries — i.e., those which have been slow to modernise — the main point here is that it lies at the centre of capital's restructuring around science and technology (as Mike Duncan has also argued in this volume).

The issue, then, is not modernity as such, but — as it has always been — the extraction of surplus value from living labour. As British industry re-tools, the microchip will tend to re-create sweatshop conditions anew. The capitalists can remain confident that people displaced by new technologies will be willing to accept such conditions, unhindered by 'trade union attitudes', for as long as the labour movement continues to oppose the consequences on capital's own terrain.

Acknowledgements

This article is based upon the paper I presented at the 1978 CSE Conference. I developed that paper through the CSE Labour Process Historians Group (later the Labour Process/Left Strategy Group), in an attempt to synthesize my earlier analyses around Grunwick. These included my articles in RSJ and the Leveller, as well as a talk which I gave on 14 March 1978 in the RSJ Seminar series.

Of the many comrades who gave me helpful comments on this article and earlier versions, I would like to thank in particular Michael Chanan, Tony Bloor, Kevin McDonnell, Frank Jabobs, Bob Young, Luke Hodgkin, Maureen McNeil, members of the Grunwick Strike Committee (especially Kamlesh Gandhi), the Institute of Race Relations (especially A. Sivanandan, Jenny Bourne and Danny Reilly), and the Poster Collective. It was in particular the Poster Collective's involvement in the strike which opened up the space for political debate both during and afterwards, and which in turn enabled me to sharpen my ideas through further discussions with the Collective.

Notes

1 ACAS was rooted conceptually in the 1969 Donovan Report, which set out a bipartisan consensus approach to industrial relations. As ACAS said in its 1977 *Annual Report*, perhaps by then somewhat defensively, 'It has been the policy of all the main parties in Parliament for very many years to regard collective bargaining as a normal method of settling terms and conditions of employment.' (p.39)

Though a precursor of ACAS had been created by the Heath government (1970–74), it was not offered as a potential weapon for trade unionism until after Heath's fall, as working class solidarity destroyed the Industrial Relations Act in practice. The new Labour government, under the Employment Protection Act (EPA) of 1975, charged ACAS with 'the general duty of promoting the improvement of industrial relations, and in particular of encouraging the extension of collective bargaining...' But ACAS was not empowered to enforce its own recommendations, as was the cautious intention of both the Trades Union Congress (TUC) and the Confederation of British Industries (CBI), whose representatives (along with academics) constituted ACAS' governing Council.

If ACAS fails to resolve a union recognition dispute through conciliation under the EPA section 11, then under section 12 it must 'consult all parties who it considers will be affected by the outcome' before reporting its recommendations. Section 14 in particular requires that it 'shall ascertain the opinions of workers to whom the issue relates'. Grunwick was able to sabotage ACAS' mission by stalling on an ACAS ballot of the remaining workforce on grounds that the already-balloted strikers were no longer his employees. Grunwick's legal action against ACAS for balloting the strikers ultimately failed to disqualify them, but the legal decision still left ACAS on weak grounds for recommending union recognition without first balloting the workers still employed there. Catch 22?

2 These lines of investigation were originally suggested by Michael Chanan in the paper he presented on 'The Political Economy of Grunwick' at the July 1977 CSE Annual Conference. He develops the broader theme of time-economy in his *Wedge* article.

3 Of course, the relatively privileged positions from which these Asians were ousted were themselves products of the way that British colonialism directed the development of those East African countries (Uganda, Kenya and Tanzania). After independence, anti-imperialist sentiment got directed at those whom imperialism had used to mediate between themselves and the African masses.

4 See Beechey for an analysis of why female labour power has a lower value and is sold at a price below its value. 'One reason why one might expect capital in general to tend towards employing married women in preference to semi-proletarianised workers is that when the married woman enters into wage labour the family is entirely dependent upon the wage — indeed, two wages, in the case of the typical nuclear household.' (p.58)

5 Amrit Wilson, 'Asian Women', a talk given on 18 April 1978 at the Institute of Race Relations, London. See her book and *Race & Class* article, as well as related articles by other writers in *Mukti*.

6 A similar impulse was manifested in the firemen's strike the following autumn (1977). Occupying fire stations was not merely a strike tactic to prevent scabbing; the firemen did not totally withdraw their labour but refused its management by the Government. They decided whether and how to respond to each fire, in some cases explicitly distinguishing between those fires which endangered only property and those which endangered human life.

7 *Race & Class* Winter 1978, p.288.

8 *Race & Class*, Summer 1977, 72.

9 *ibid.*, 73. This subtle '*management* of racism' became necessary because of the unique form of internal colony in Britain, as analysed by Sivanandan, 1976. Most Western European countries developed a contract labour system and nationality laws which 'prevented the integration of migrant labour into the indigenous proletariat and thereby mediated class conflict'. (p.351) By contrast Britain developed its internal colony under its special Commonwealth relationship (before 1973) offering British nationality to immigrants, especially to the East African Asians.

10 Yasu Patel speaking in an interview, *CARF* 3; excerpted in *Race & Class* Winter 1978, 293–4.

11 The Social Contract strategy for using the Grunwick strike resonates closely with the broader theory of race relations management formulated by John Rex's school of political sociology. Rex calls upon the state and the trade unions to help integrate blacks into the mainstream of British working class politics lest they turn instead to 'militant black movements' refusing to identify with British society altogether, 'preferring to adopt a pariah position, being in the society and not of it, and developing their own insulated social system'. For Rex the antidote is what he calls 'British class politics', by which he means a struggle not over the social relations of production but over differential access to markets. (See Bourne, pp.348–9)

It is within that notion of class, treating society as a 'system of class positions', that Rex needs to assume that the white working class has already gone far in overcoming class domination by using the rights which enhance equal access to markets and hence to class positions:

> . . . the worker in the welfare state has a set of rights as a citizen which are more important to him than those which derive purely from class membership.

While that reassurance is in itself unremarkable (as social democratic ideology goes), it provides the perspective for Rex's imaginative leap. He extends the market model from the welfare state proper to the area of employment, treated as a marketplace of labour possibly entailing unequal access (discrimination). Hence the workplace offers opportunity for integration through more equal access ('upward mobility'), as well as the danger of instability:

> Our politics are based on class conflict [and] we have a class struggle over employment, wages and housing . . . all this constitutes a *manageable* system compared with the rift which might be opened up by the continued suppression of Asian and West Indians into an "underclass' position.
> —John Rex, quoted in Gilroy, pp.59–60.

12 Manchester M–L Group pamphlet, pp.6, 10.

13 See Holloway & Picciotto for the separation of production relations into economic and political forms, as an historical and contradictory tendency of capitalism: 'It has ever been characteristic of reformism that it emphasizes the reality and not the illusion of the fragmentation of social relations that it accepts as given the fetishization of class struggle into distinct economic and political channels.' (p.81) See Clarke for its basis in a technicist mystification of production which makes bourgeois relations of production 'appear as relations already inscribed in the technical structure of the material production process by ascribing revenues to factors of production', representing labour as one technical factor among many. (p.4)

Also see Elson for why 'exploitation appears to take two separate forms: "unfair" money wages or prices, and/or arduous work with long hours and poor conditions'. Her article explains how Marx's theory of value 'enables us to analyze capitalist exploitation in a way that overcomes the fragmentation of the experience of that exploitation'. (p.171)

14 A less obvious factor in the Government's quandry over Grunwick was its 'industrial strategy' which included encouragements to small industries to remain in the inner city areas. Even if the Social Contract was to provide class peace as an added incentive, in the case of Grunwick it turned out to lead to management threats of shutdown and/or relocation. Hence the unanticipated contradiction: a provider of jobs acting as a political renegade, and so preventing the capitalist state from stabilizing capital.

15 An article in *Race Today* (Nov/Dec 1977) reproduced the front page of the 3 Nov 1977 issue of *Socialist Challenge* as a symptom of the problem. This front page had played up (pictorially) the way that the Grunwick Strike Committee misguidedly attempted to build for the November 7 'Day of Reckoning' as continuing the grand tradition of British trade unionism — which has almost consistently let down Asian workers. *Race Today* counterposed a more insular strategy of appealing to the tradition of Asian working class militancy in Britain, largely outside the trade unions. But the magazine's criticisms could lead us instead to a less insular strategy of using Grunwick as a way of breaking the hold of state-corporatist trade unionism over the white working class by appealing to its impulses to supersede trade unionism, by seizing political control over the labour process.

16 In the end, the Cricklewood sorters who had selectively blacked Grunwick's mail were left defenceless. Seven UPW members and local officials were fined up to £500 by their own union! Less well known is the debacle over the hundreds of arrested pickets, who lost out on the possibility of a collective defence in the courtroom after Jack Dromey stalled on calling a public meeting for those arrested.

17 The crucial distinction between *exposing* and *challenging* betrayals — between morally denouncing them and organizationally preventing them — lies at the heart of the debate over Grunwick in the *Leveller*. Leo Panitch draws a relevant lesson from the entire period of the Social Contract: 'In the case of the Labour Party, self-defence takes place *at the expense* of the hegemonic development of the working class.' (p.59) This lesson applies as well to the much broader 'labour movement-ism' which appeals to delegates and representatives to defend our rights for us, thus discouraging those in struggle from grasping the possibilities for taking control over the situation, for making the labour process their own. Or, as Karl Marx warned, 'Whomever one seeks to persuade, one acknowledges as master of the situation.' (*18th Brumaire*, p.74)

18 The complementary capital-intensive and labour-intensive stages found at Grunwick appear on an international level in the

microelectronics industry. The production of silicon chips features the intensive exploitation of Southeast Asian immigrants in the fiercely competitive Silicon Valley as well as in the Free Trade Zones maintained by SE Asian ruling elites to compete for foreign investment.

See Sivanandan, 1979, on Free Trade Zones; see Mike Duncan in this volume on Silicon Valley.

19 Guy Debord, *Society of the Spectacle*.

20 Paul Simon, 'Kodachrome'.

21 In the British Steel strike of early 1980, Lord Denning attempted to re-define the law by declaring action in the private sector 'illegal'. Steel union leader Bill Sirs successfully called upon his members to comply with Denning's order until the Court of Appeal declared it groundless — thereby helping to strengthen the authority of both the law and the official unions to determine the permissible forms of struggle.

Furthermore, Sirs defended action in the private sector by arguing that it was in fact 'primary' — thereby accepting capital's definition of action as either primary/secondary. Arthur Scargill, by contrast, refused to accept such terms when he insisted upon calling it just plain 'effective' action; this was more to the point but still evaded the problem of its political illegitimacy as a weapon of industrial disputes.

This is of course a problem for us all; as I was re-writing passages from my 1978 CSE paper on Grunwick, I found myself subconsciously changing my original phrase 'selective blacking' to 'secondary blacking'. How do we mount a counter-attack on capital's definitions of our reality?

22 On the self-policing role of the trade unions, see 'A Tory "Social Contract"?', my paper for the 1980 CSE conference.

Bibliography

John Andrews, 'Guardian reporters assess labour relations in the film processing industry', *Guardian* 23 June 1977.

ACAS, *Annual Report 1977*, 1978.

Veronica Beechey, 'Some Notes on Female Wage Labour in Capitalist Production', *Capital & Class* 3 (Autumn 1977), 45–66.

Jenny Bourne, 'Cheerleaders and Ombudsmen: the sociology of race relations in Britain', *Race & Class* XXI, 4 (Spring 1980), 331–52.

CARF (Campaign Against Racism & Fascism) no.3, 'Grunwick Women', interview with Jayaben Desai and Yasu Patel, p.6; excerpted in *Race & Class* Winter 1978, pp.293–4.

Michael Chanan, 'The Kodak Shift: Electronics, Media and Multinationals', *Wedge* 2 (Spring 1978), 32–8.

Simon Clarke, 'Marxism, Sociology and Poulantzas' Theory of the State', *Capital & Class* 2 (Summer 1977), 1–31.

Guy Debord, *Society of the Spectacle*, Detroit, Black & Red pb, 1970; excerpted as 'The Proletariat as Subject and as Representation' in *Leaving the 20th Century*, London, Free Fall Publications pb, 1974, pp.110–126.

Jack Dromey and Graham Taylor, *Grunwick: The Workers' Story*, Lawrence & Wishart pb, 1978.

Diane Elson, 'Marx's Value Theory of Labour', in Elson (ed.), *Value: The Representation of Labour in Capitalism*, CSE Books pb, 1979, pp.115–180.

Paul Gilroy, 'Managing the "Underclass" ', *Race & Class* XXII, 1 (Summer 1980), 47–62.

John Holloway & Sol Picciotto, 'Capital, Crisis and the State', *Capital & Class* 2 (Summer 1977), 76–101.

Les Levidow, 'Grunwick as Technology and Class Struggle', *Radical Science Journal* 6/7 (1978), 119–128.

Les Levidow, 'Grunwick Collapse', *Leveller* 12 (February 1978), 12; reply in no.13 (March 1978), 22; debate in no. 14 (April 1978), 10.

Les Levidow, 'The Social Contract as Reformism: Grunwick as a Case Study', *CSE Conference Papers*, 1978, 266–277.

Les Levidow, 'A Tory "Social Contract"?', *CSE Conference Papers*, 1980, 126–30; updated version in *Chartist* 82 (Nov/Dec 1980), pp.19-22

Manchester Marxist-Leninist Group, *The Grunwick Strike: Struggle, Solidarity, Betrayal and Defeat*, spring 1978 (available from 109 Oxford Road, Manchester 1).

Karl Marx, *The Eighteenth Brumaire of Louis Bonaparte* (1852), Moscow, Progress pb, 1954.

Mukti: Paper of Asian Militants. Jayaben Desai interviewed by S. Sinna, 'Fighting for Dignity and Respect', *Mukti 1* (December 1977), 6–8. Yashmin Khan, 'The Social Basis of Arranged Marriages', *Mukti* 2 (April 1978), 23–27.

Leo Panitch, 'Socialists and the Labour Party: A Reappraisal', *Socialist Register* 1979, Merlin Press pb, 51–74.

Poster Collective, *Women in Struggle*, 1978, pp.36–9.

Race & Class XIX, 1 (Summer 1977), 69–73; *Race & Class* XIX, 3 (Winter 1978), 289–94; XXI, 4 (Spring 1980), 331–52; XXII, 1 (Summer 1980), 47–62.

Race Today April/May 1977, 52–54; Sept/Oct 1977, 122; Nov/Dec 1977, 151–4; Jan 1978, 22–3; Aug/Sept 1979, 79–85; Nov/Dec 1979, 103–5.

Joe Rogaly, *Grunwick*, Penguin pb, 1977.

Lord Scarman, *Report of a Court of Inquiry... into a Dispute Between Grunwick... and Apex*, HMSO Cmnd 6922, 1977 (excerpted in Rogaly).

Paul Simon, 'Kodachrome', *There Goes Rhymin' Simon*, CBS 569035 (1973).

A Sivanandan, 'Imperialism and Disorganic Development in the Silicon Age', *Race & Class* XXI, 2 (Autumn 1979), 111–26.

Socialist Challenge pamphlet, *The Battle of Grunwick*, October 1977.

Spare Rib 52 (Nov 1976), 10–13; 54 (Jan 1977), 18; 61 (Aug 1977), 6–8, 48; 66 (Jan 1976), 40.

Sunday Times 23 August 1977.

The Times editorial 30 June 1977.

TUC Report 1976, p.466.

George Ward, *Fort Grunwick*, Temple Smith pb, 1977.

Which? June 1972, 172–6; June 1977, 308–11.

Amrit Wilson, 'A Burning Fever: the Isolation of Asian Women in Britain', *Race & Class* XX, 2 (Autumn 1978), 129–42.

Amrit Wilson, *Finding a Voice: Asian Women in Britain*, Virago pb, 1978, especially pp.59–69.

Microelectronics: Five Areas of Subordination

MIKE DUNCAN

Introduction

Capitalist science and technology have been built on the ability to repeat action, predict its outcome and quantify its result. In the constant battle between labour and capital over the production, realisation and distribution of surplus value, these three foundations have become politically charged weapons wielded by capital in its never-ending quest for total control over labour. If scientific practice is the underlying methodology then technology is the cutting edge and, in the current period of deep capitalist crisis, the new technology of microelectronics provides capital with a new blade with which to cut into labour.

Up until now most attention has been given to the effects of microelectronics on employment levels in the factory and the office. There have been various attempts to quantify just how much labour will be ejected in the production, circulation and distribution processes by microprocessor-controlled machinery.[1] If we use bourgeois forms of econometrics, then our attention is drawn only to the quantitive changes within capitalism's current strategy of restructuring. However, the number of workers forced out of jobs, their position in the class structure as well as their race and sex, and the total number of new members of the class who will not be able to find work will depend on the relative strengths of labour and capital in their battle for control.

We are going to look at five key locations of this battle, locations where capital is already using, or hopes to use, microelectronics as a weapon. In this way we will look at the dynamic aspects of the form and content of class struggle and so undermine the key concept of progress as an inevitable process unshaped by human action.

We begin by looking at the production of microelectronic chips themselves. This is because the very existence of this technology is dependent upon a divided, atomised and subordinated working class forced to labour at a furious pace in order to fulfill the plans of designers and engineers who have used the ability to repeat, predict and quantify electronic phenomena as the basis for the technology.

But capital has been less successful in its attempts to sustain control over labour in the production of software to instruct hardware chips what to do — this is the second area we analyse. Capitalist managers do not yet fully understand the processes of software writing, just as they have failed to control other crucial areas of clerical work, and they are therefore not yet able to seize control of the experience and judgement of software workers. The tacit knowledge of programmers and systems analysts is shielded from management by the continuation of craft practices. Although division of labour, real subordination to machinery, methods of quantification and supervision have all been tried to some degree or other, capital has not yet achieved a sustained initiative over labour in these production processes.

Functions in work that capitalist management cannot break down into quantifiable, predictable and repeated actions are classified as unstructured. One of the biggest areas of unstructured work, which grew most rapidly in the last phase of capitalist expansion spanning from 1940 to 1974, is office work. The way capital hopes to use microelectronics to structure office work is examined in our third section, which covers office automation. The office not only aids the circulation of capital but also recomposes the functions that have been split off from the labour process by mass production techniques, and clerical work co-ordinates the growing supervision of the reproduction of labour undertaken by the state. Women play key roles in the office both as general clerical labour with typing skills and as the secretaries of managers and professionals. As secretaries they hold a lot of information and use their experience and judgement not only to maximise the efficiency of their 'principal', the male manager, but also to create time and space within their own jobs for their own use. The automation of offices, possible with microelectronic technology, is aimed at squeezing this time and space to a minimum and finally eliminating the role altogether. The functions that are already structured in a secretary's work will be broken off and automated and those that are now unstructured will be subject to rigorous systems analysis in a sustained attempt to decompose them into structured tasks. As this process gets underway, driven partly by the multinationals' insatiable demand for new markets, then the work of managers and professionals themselves will be subject to the same scrutiny.

Underground coal miners in the UK are a group of skilled workers who exercise a considerable level of control over their work rates and conditions — the automation they face is discussed in our fourth employment analysis. The UK's National Coal Board is planning to place microelectronics-based equipment between the worker and the extraction process, breaking the present power relationship between miner and manager in which the manager cannot gain the upper hand, and can recast the relationship only by installing machinery to which the miner will be subordinate. Control over production will be shifted to the surface where management will oversee the whole process at a computer terminal.

Finally we look briefly at an area where the development of microelectronics offers the potential for creating new jobs with the expansion of a new industry, increasingly dependent on the new technology. The UK radio communications, radar and navigational aids industry has responded to expansion with the adoption of more advanced microelectronics-based computer aided design and manufacture systems instead of with training programmes to reskill workers. A chronic shortage of skilled workers is being met by the substitution of capital equipment instead of by an expansion of labour in the industry.

Our review is not meant to be all-embracing. Instead we throw light on certain areas, concentrating in one place on the composition of labour, and in another on the strategies of capital. All this should be read in the light of our understanding that neither labour nor capital has a hegemony of control and that the very shape of an industry and the labour process that goes on within it are determined by the form and content of class struggle — a struggle which capital can never win once and for all.

Microelectronic Chip Production

The production of microelectronic chips — microprocessors and micromemories — is dominated by a handful of companies that have production plants in the United States, Western Europe, Japan and the capitalist enclaves of the Far East. These

FREEING LABOUR...

BRITISH MANAGERS UNDER THE SHACKLES IMPOSED ON THEM BY GENERATIONS OF TRADE UNION DOMINATION MAY HAVE TO RELEARN WHAT CAN BE ACHIEVED WITH FREEDOM OF CHOOSING THE MOST EFFECTIVE DEPLOYMENT OF LABOUR. WORKERS WILLING AND PERMITTED TO DO ANYTHING MAY ALSO PROVE MORE VALUABLE THAN MEN LIMITED BY THEIR CRAFT UNIONS THAN CAN NOW BE FORSEEN. Prof.Hayek,monetarist economist.

Does your labour supply measure up to requirements?

g.

Double output per man

At present it appears that

... THE OPERATOR WILL NOT HAVE TO REMEMBER COMPLICATED WORK SEQUENCES, AS SOON AS THE MACHINE HAS BEEN USED ONCE IN THE PRESCRIBED MANNER THE PROGRAMME IS STORED IN RANDOM ACCESS MEMORY. FROM THEN ON THE OPERATOR (OR ANY OPERATOR) SIMPLY HAS TO GUIDE THE MATERIAL AND HOLD THE TREADLE DOWN.

... THE MACHINE MEANS THAT INSTEAD OF GIVING LESSONS TO OPERATORS THE THE SEWING MACHINES THEMSELVES WILL RECEIVE TRAINING, COUNTERING STAFF TURNING AND TRAINING, COST PROBLEMS.

Singer Centurian Programmable sewing machine. Financial Review 8 - 8 - 1978

TO SET THE WORKER FREE?

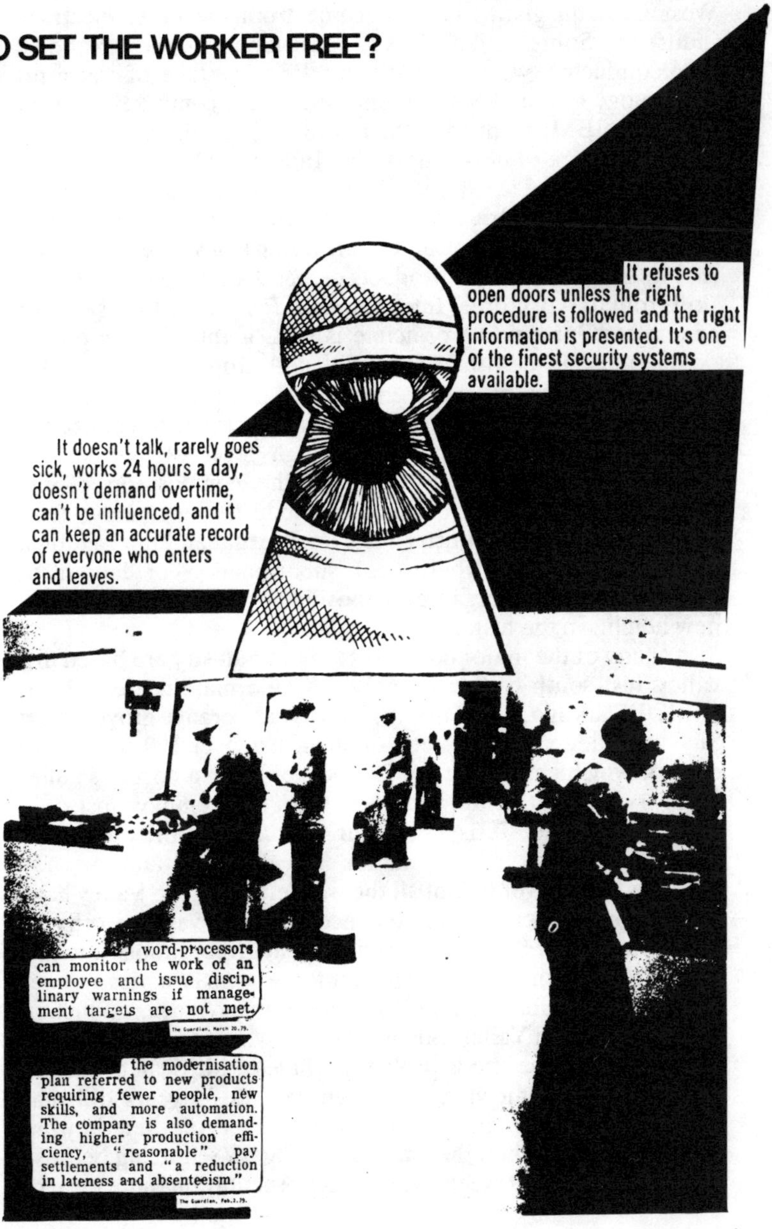

companies have a variety of backgrounds. Some of them, like the West German giant Siemens, come from the older electrical industry. Some, like Texas Instruments and National Semiconductor, were founded to take advantage of transistor technology in the 1950s. Others were in the computer industry and, like IBM, went into their own chip production. Finally, there is a block of newcomers, like Intel and Mostek, who were especially formed to exploit this new microelectronic technology.

The microelectric industry, of which US-based firms control 63%, spends a lot of time and effort trying to give the impression that its high technology products demand an increasing level of skill from its workforce. Intel's *Annual Report* for 1978 says, for example: 'Our guiding principle is that, if Intel is to grow, its people must grow with its capabilities. A growth company must be, among other things, an educational institution.'[2] Intel, and the other manufacturers, say there is an immediate connection between high technology and a high level of skill in the workforce. We shall now look at the composition of the workforce in the microelectronics industry and show that, instead of skill, the industry rests on a sexual and international division of labour where real subordination of labour to production and the organic composition of capital have reached new heights in the history of capitalism.

Many of the major companies in this industry are based in a valley just south of San Francisco, California, an area which, until 20 years ago, was entirely given over to orange groves. Now Silicon Valley, as it has been dubbed, has over 100 companies each of which has more than 250 workers, and a host of smaller firms trying to start up. Between them they employ just over 180,000 workers. It is here that the vast majority of chips available on the world market are designed and actually produced. Just over half of all the workers in Silicon Valley have jobs in production or maintenance, and the vast majority of these, close to 80,000, are women. Recent investigations into this almost totally non-unionised labour force by an American trades union have produced a telling picture of the average production worker in Silicon Valley. She is a woman whose second language is English because she is probably Chinese, Mexican or Puerto Rican. She is somewhere between 18 and 30 years old, has dependent children and has to pay for their care when they are not at school or when they are too young for schooling because she works a 40 hour week. She is most probably on some form of

FREE TRADE ZONES

One worker working 1 hour produces enough to pay the wages of 10 workers working 1 shift plus all the cost of materials and transport.

Manager of plant in Malaysia

THE GOVERNMENT COULD NOT HELP THE PEOPLE IF THEY REFUSE TO REALISE THE IMPORTANCE OF A BETTER ECONOMY AND TO BE MORE RESPONSIBLE... WORKERS MUST UPHOLD THEIR DIGNITY AND NOT CAUSE PROBLEMS THAT WOULD SCARE AWAY FOREIGN INVESTORS. THEY SHOULD INSTEAD BE MORE PRODUCTIVE SO THAT GOVERNMENT EFFORTS TO ATTRACT INVESTORS WOULD BE SUCCESSFUL.

Malaysian Deputy Prime Minister,

Mahatir Mohamed in article, "Why We Woo Foreign Investment"

Answering wage demands in Malaysia

<u>MALAYSIA - MESSAGE TO THE MULTI-NATIONALS</u>

- NO MINIMUM WAGE LAW COVERING MANUFACTURING INDUSTRY.
- 'LOW COST' RELIABLE AND 'RESPONSIBLE' LABOUR AT $1:20 PER DAY.
- AN EMPLOYEE MAY BE SUMMARILY DISMISSED.
- ONLY SEVEN DAYS ANNUAL LEAVE FOR WORKERS.
- SAFEGUARD OF 'PIONEER' INDUSTRIES AGAINST UNREASONABLE DEMANDS OF TRADE UNIONS SUCH AS ACTION OVER SACKING.
- ANY YOUNG PERSON OVER SIXTEEN MAY BE EMPLOYED TO WORK WITHOUT ANY RESTRICTION.
- WORK STOPPAGES ARE FEW AND FAR BETWEEN. ONE OF THE MAIN FEATURES OF MALAYSIAN INDUSTRIAL RELATIONS LAW IS THE PROHIBITION OF STRIKES ONCE THE GOVERNMENT HAS INTERVENED.

Quoted in Rachael Grossman's paper, "Woman's Place in the Intigrated Circuit".

welfare because her pay, just over $5,000 a year, is either at or just above the legal minimum.

The skills that she has within the production process are probably only recognised in some title which will be embossed on the plastic badge attached to her white coat. For supervising other workers, however, she will get increases in her salary. While on the production lines, which are contained in air-conditioned cabins, she will be exposed to chemicals of all types used in the different cleaning and manufacturing stages. Side by side with these organic solvents, which can irritate the skin, cause rashes, dizziness, nausea and vomiting, she will be working with highly corrosive acids. All the production processes, from the first cutting of a sausage of silicon into wafers, right through to the final cutting of the completed chip, will be monitored by computer systems. The computers provide a basic pace for the work to which production workers have to respond with an equal pace in their own labour. The design and management functions within these plants are mostly undertaken by men. As well as depending on a sexual division of labour within the metropolitan countries, capital in the microelectronic industry has also used an international division of labour. By 1963 the major companies in the electronics industry began to hunt the capitalist world's labour market in search of the cheapest labour which could be used to assemble electronic devices into electronic and electrical systems. Fairchild established a factory in the British colony of Hong Kong in 1963, and Taiwan, South Korea and Singapore became firm favourites for overseas assembly plants in the 1960s. In the early 1970s today's microelectronic industry was formed when it became possible to put more than 1,000 transistors onto a single chip. This new level of integration led to a new wave of expansion into Malaysia, Thailand, the Philippines and Indonesia from 1972 to 1974. The new wave formed a working class which had no previous experience of wage-labour. By 1978, for instance, Malaysia had 69 electronic plants, twenty of them owned by US corporations. The country had become, according to a US Department of Commerce report, 'one of the major manufacturers of semiconductor products in the world'.[3] The value of production in that one country grew from $382.9 million in 1976 to $500 million in 1977, an increase of 37%. 'At the present time', the Department of Commerce report concludes, 'some well established electronics companies such as Monsanto, Motorola, RCA, Texas Instruments, have gone into third shift

"We hire girls because they are easier to control . . ."

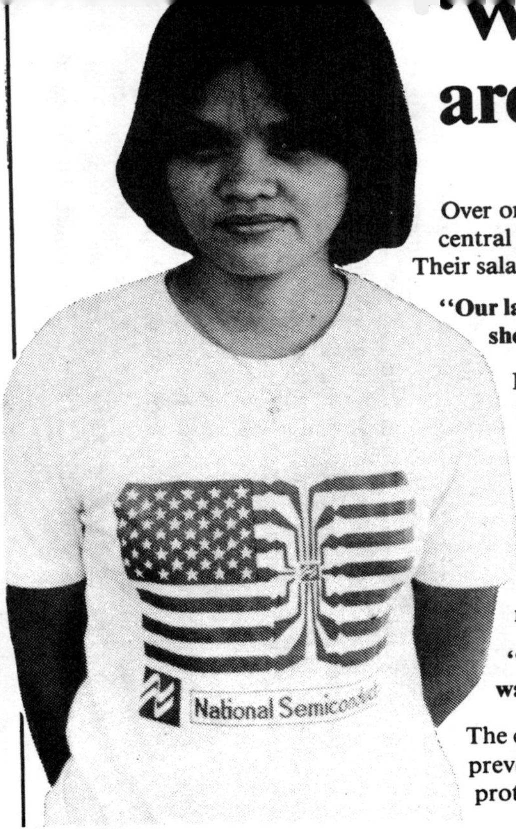

Over one million S.E. Asian women now work for U.S. corporations, forming the central link in assembly lines that stretch from the U.S. to Asia and back again. Their salaries are as low as 80¢ a day.

"Our last winner of the company beauty contest spent $40 on her evening gown, but she made so many slits to show her legs, that she can't wear the dress anymore."

New forms of personnel control which manipulate traditional concepts of femininity, passivity, and sexuality are now being implemented by American electronics companies in S.E. Asia.

"Hey grandma! How do you like your new glasses?"

Electronics workers in Hong Kong are called "grandma" after they reach the age of 25, since by this time they need eyeglasses. Asian women work in these factories only 3 to 4 years before they suffer severe eye damage. As a result, they often lose their jobs.

"Intel doesn't believe in unions, because we believe in finding out what workers want. Our company conducts a twice yearly attitude survey."

The corporate strategy in Asia is to divert independent worker organizing, and to prevent employees from attaining their legal rights even under the limited protection of S.E. Asian regimes.

per day production.'

In these plants in South East Asia the vast majority of workers are women who have had no formal training in microelectronics. They, like their sisters in Silicon Valley, are not unionised, have no formal recognition of their skills and are the target for a constant barrage of sexist propaganda orchestrated by the personnel department of the firm aimed at reinforcing subservience and loyalty to the company.[4] According to the Director of the Institute of Race Relations in the UK, this exploitation has meant that production, 'to the masses of these countries, is production without purpose, except to stay alive; massive immiseration accompanied by a wholesale attack on the values, relationships, gods that made such immiseration bearable; rulers who rule not for their own people but for someone else — a development that makes no sense, has no bearing on their lives, is disorganic.'[5]

Wherever microelectronic plants are located, therefore, the multinational companies that own them are looking for a universal profile of labour to fit into the production process. The scant recognition these workers get for the skills they use in production is part of a sexism which ensures that their skill is never recognised by capital in anything but title and that they are given the most mean wage increases. At the same time this labour force has remained largely docile when it comes to the more traditional organised struggles of the working class for more pay, less work and better conditions. Instead, since the late 1960s, capital has retained an initiative over these workers by rapid changes in the production processes, increasing rates of automation in key areas and expansion into areas of high unemployment or precapitalist production where client states can deliver a submissive labour force.

It is on the backs of these workers in the metropolitan countries and at the periphery that the microlectronic industry has developed. The sustained initiative that capital has retained in this industry has driven it to the point where such a massive injection of capital is needed to start up and break into the market that even large corporations cannot do it.

To establish a production line today with the capacity to produce about 2½ million chips a year with present methods would cost an estimated $8,500,000 for the capital equipment. From this line chips selling at about $8 each would gross annual sales of about $20,000,000. The typical product would be a memory chip capable of storing 16,000 binary digits of data.[6]

This high entry cost can be compared with the price that a leading company had to pay in the late 1960s and early 1970s to get into the business. Between 1969 and 1972 Intel, which then had 200 workers at the beginning of its investment programme, spent $3,850,000 establishing its plant and equipment in the US for production of microelectronic memories and processors.[7] The plant Intel established was soon able to produce memories which could store 1,000 binary digits and from this investment Intel was able to generate $23,000,000 worth of sales in 1972 with a workforce of 500. Looking to the future, it has been estimated that to establish a plant in 1982 will cost about $18,000,000 in capital expenditure to make storage devices holding 64,000 binary digits and, in 1985, for chips that can store 256,000 binary digits, the grub stake will have risen to $40,000,000.[8] These barriers to entry have been created not just by the technical demands of the production processes but also by capital's continuous need to retain its initiative by making the very production process itself obsolete. 'Technological advance' within the industry is therefore rapid and aimed in two directions at once — keeping out other capitals from the industry and keeping the workforce docile. As it will have little time to organise, its position in any one production process is always under threat by the next 'technical advance'.

Male and female technologists and scientists working for these firms imagine themselves to be involved in purely technical work to which they can relate with a spirit of professionalism and scientific adventure.[9] These companies spend millions of dollars a year on research and development — Intel spent $41 million in 1978 alone. These resources are channelled into solving problems of how to design and manufacture chips that will consume less power, give off less heat, store more data, process that data faster and all at a lower cost. Arguments about the application of microelectronics do not move these professionals to examine their own labour processes and their relations with capital. But what they have never been challenged on is the essential function they fulfill — a function which allows capital to make the production processes obsolete every four years and therefore to decompose the skills of the labour force to the level we had described. These technological and scientific workers therefore make a direct contribution to the continual victory of capital over production workers — in this way their work has a class content irrespective of their individual intentions and aspirations. What their own fate may be as 'professional' workers we will see when looking at office automation. The fact that such a high technology industry, spending so much on research and development, with such rapid 'technical advance', is based on a foundation of workers who have so little control of their own labour and are subject to such high rates of exploitation is a central contradiction of modern capitalism. When we look at the production of programmes to run on these microelectronic devices we will see another contradiction of capitalism, also based in the labour process.

Software Production

No microelectronic device can be used for any application, whether it be the control of door chimes or the monitoring of levels of stock, unless a program of instructions is drawn up for it. Microelectronic processors have to be instructed what to do. And the programs, called software, have to be loaded into the microelectronic memory so that the electronic device can work through the instructions at its speed of calculation and storage retrieval.

In contrast to the production of hardware, the production of software is, at this moment, an area where capital has been unable to achieve a sustained initiative over labour and there is therefore, from the point of view of capitalist management, a crisis of software productivity. Having bought the labour of programs and systems analyists as a commodity on the market, capital is, at the moment, unable to raise the real subordination of this labour to capital to anything like the level it has achieved in hardware production processes.

This insubordination of labour within the software production process has many expressions, several perceived clearly by capitalist management. The relatively high cost of software production is put down by capitalist management to the fact that designing and writing programs is a labour intensive process. This means that the essential tasks in the production of software are largely executed *under the control* of labour instead of being executed under the control of capital. No one in hardware manufacture would conceive of the microelectronic device they are working to produce as 'their chip'. Yet in software production the handicraft relationship between labour and its product exists to the point where a programmer frequently talks of a program he is working on as 'my program'. This relationship reflects directly the lack of control which capital has over the process of production, and a key part of capital's attempts to regain some initiative in the struggle around the labour process is aimed at it. The weapon capital is using is the division of labour. A recent IBM publication for managers of software production departments outlined new methods of organisation resting on a division of tasks to create the special function of librarian. IBM's proposed job definition for the librarian is:

> 'A Librarian is responsible for picking up all computer output, good or bad, and filing it in the noteboooks and archives of the development support library where it becomes part of the public record. By contrast in traditional programming operations the bad runs go into the waste basket, often destroying information of latent value, but certainly destroying information about errors of carelessness. This identification of all program data and computer runs as public assets, not private property, is a key principle in team operations.'

The product, and all that labour does to produce it, can not, from capital's point of view, remain the 'private property' of labour — it must become the private property of the corporation. Output from 'bad' runs, attempts to get the program running that have not worked to specification, is positive proof to capitalist management of errors of carelessness. Unless this is out in the open, capital will have little chance to control the productive process of software production because it will be unable to monitor it.[10] When faced with such information from the notebooks and archives the programmer will be at a distinct disadvantage in project review and career review meetings with his or her management because s/he will be unable to appeal to a higher technical knowledge of software as management has documented proof of his or her 'inadequacy'. Once again the product of labour will confront labour itself in the battle between labour and capital.

This battle over the control of the product goes hand in hand with attempts to quantify the process of software production in order to measure performance of labour against time and costs. By the nature of hardware production this quantification is available and used by capitalist management not only because of its continued initiative over labour but also because methods of engineering production quantification have been developed in earlier stages of capitalism. But, to date, software has not proved as easy to quantify, not only because of the continued control of labour, but also because the production process itself is relatively new in capitalist history. This is a process in which the ability to handle uncertainty and change are still paramount qualities of the labour. Despite time and motion studies and the inappropriate application of crude productivity indicators, capital has yet to devise methods which adequately reflect the unique characteristics of the labour of the mind as distinct to the labour of the hand. A leading international consultant in the management of software projects amply expressed the dilemma capital faces:

> It is well recognised in business that control over our activity is proportional to our ability to measure . . . I find it rather interesting to observe, as a neutral observer (sic), that the developments in the metrics area do not come from the computer science academics. They come largely from the trained engineers who are applying their engineering concepts — "everything of practical importance can be measured somehow" — to their research and practical work. Very little work is coming from universities. Most of it is being developed by organisations committed to mastering the very large problems of highly reliable and large software systems . . . Management will not pay the bill for our lack of control over the technology when they see an alternative, and the alternatives are here.

He is, frankly, optimistic, from capital's point of view, when he says that adequate levers of control exist. But one such lever was outlined in the monthly journal most frequently read by management in the computer industry. *Datamation* held out the experience of a US computer service company providing processing power and software for the telecommunication and electronics industries as a model to others. The senior management of the firm began a three year scheme, the objectives of which were:

1. Establish performance objectives.
2. Record and report total performance.
3. Provide a standard basis for project comparison to aid in determining future requirements and in specifying where costs should be reduced.
4. Measure and compare performance of one programming performance with another.
5. Foster a competitive spirit that could be an incentive for

improving service and performance.

6 Demonstrate to others that data processing management was concerned about increasing productivity while improving the overall quality of its products and services.

The system that the firm developed involved counting the lines of 'error free' code written by programmers, measuring overtime, determining what labour was spent in servicing various computer devices and measuring how fast programmers could 'fix' bugs in programs. This detailed level of monitoring squeezed out the time and space that workers had created within their jobs. Without any substantial changes in work methods, productivity went up in the 14 data centres where nearly 1,000 software workers were employed.

For instance, even before the management began to tackle the 'bottlenecks' in production revealed by the surveillance and data collection projects, the number of 'errors' per line of code written by programmers dropped by 59%; the amount of a later adaptation and maintenance that each 1,000 lines of written code needed dropped by 47%; and the overtime worked by the 1,000 staff fell by 56%.

The initiative of senior management had dragged the process of software production into the open, out from the minds and closed social relations of one programmer to another, out to a point where management could publicly examine the process. In conclusion a senior manager on the project said:

> As many functions as possible must be removed from the 'art' category of management. A measurement system is simply another tool which can be used by an alert management to increase productivity and quality, and to make the overall job of managing data processing easier.

In all, the three essential characteristics of capitalist science method have been imposed on this work: it has become *repeatable*, management can *predict* its outcome and *quantify* the result.

This initiative by one individual firm seems, from the point of view of the capitalist management, to have been successful. But many such attacks on labour's control of the software production process have looked promising when started but have soon got bogged down as unofficial forms of resistance by labour developed. As this struggle continues capital often uses the productivity it has extracted from hardware production in its attempts to outflank software workers. There is therefore an increasing tendency by management to rely on the speed and density of the new microelectronic devices to 'bury' the unproductive aspects of software.

The most dramatic example of this is the development of high level programming languages nearly 20 years ago. Before these languages, Cobol and Fortran, were produced, programs had to be written in instructions that bore a close similarity to how the actual electronics worked. These low level languages were specific for each range of computer and demanded of the programmer a high knowledge of how the internal hardware worked.

At this time the majority of programmers were graduates from an educational system that had not yet felt the rapid expansion of resources and intake which occurred in the middle 1960s. The hardware technology was based on the use of vacuum tubes and other electronic components as well as storage devices like mercury delay lines supplied by electrical manufacturers. Only six years before, in January 1954, the first commercial application of a computer had been successfully completed when the calculation of the payroll for 1,700 workers at a Lyon's bakery in West London processed the weekly details and printed the payslip for each worker in 1½ seconds.[11]

At the time of the Lyon's Electronic Office (Leo), as it was called, there was a close relationship between hardware designers and programmers. This high level of skill among software producers not only ensured them a high salary but, concurrently, provided a block to the expansion of computer applications into a wider market of clerical systems from its previously pioneering role in research and science. In the late 1950s, therefore, a movement began among the management of large computer users to develop languages where applications could be written faster.

The US Department of Defence (DoD), always prominent in the history of the computer industry, took the lead in bringing a number of major manufacturers together to define a language for commerical applications. One of the explicit aims of this move was to get a programming language that managers without a detailed knowledge of computer hardware could use to write their own programs with the 'minimum of time and programming effort'. Under DoD guidance the language designers came up with a specification for a Common Business Oriented Language (Cobol). And the initial resistance of computer manufacturers to committing their resources to developing the language for their machines was overcome when the Navy Department said a condition of future purchases would be that the ordered computer should accept Cobol code. But initial attempts to write a software package which would take Cobol code and generate from it the instructions to actually run the hardware were not, in the main, successful. Primarily this was because the management of computer manufacturers had not grasped that the productivity achieved in hardware could be partly traded for 'less efficient' software. Therefore the first packages, called compilers, the manufacturers had designed by their own software writers were often unworkable because they had set a design objective that the actual instructions to drive the hardware generated by the compiler should be as 'efficient' in the use of the hardware as the product of any average low-level programmer. Designers had taken much the same attitude to the development of another high-level language, Formula Translator (Fortran).

Only when this design objective was ditched, around 1963, did Cobol begin to work as a 'technical' solution to bottlenecks currently experienced in low-level programming. There was, concurrent with the move to Cobol, a leap in hardware performance as new computers built of transistors, not vacuum tubes, and core memories, not delay lines or other storage devices, came onto the market from 1961 onwards — another transfer of the productivity gained in hardware production to software production.

It was also about this time, and by no means a coincidence, that one of the first divisions of labour occurred in the production of software. Until now the design of the software and the writing of the programs had been one and the same task. But, hand in hand with the introduction of high level programming languages, the tasks were divided by management into what was called Systems Analysis and Programming.

Thus, in this brief period from about 1957 to about 1964, capital took an initiative against the labour in the software production process which, in the short term, was successful.

Without this initiative the explosion of computer applications in the 1960s and early 1970s would have been impossible, primarily because the productivity of software production was too low to carry such an expanded field without it absorbing millions of new workers throughout western capitalism, a move that would have forced management to expend much more in variable capital than it was willing to commit.

The low-level programmers, in the main, went in two directions — following the division of labour and becoming systems analysts, or moving over to work for the manufacturers who still wanted their programs written in low- level language for the purposes of hardware efficiency. Underneath these people a whole new generation of programmers entered the industry as Cobol and Fortran programmers who had little knowledge of computer hardware. Some 'low-level' programmers remained to fight it out and a few are still around today. They argued at the time and still argue that low-level languages are more efficient because they use the resources of the machine better, but their argument is miscast because capital was not seeking efficiency from the machine, but from the labour process of software production.

However, the initiative that capital had gained in software production had in the main replaced a group of workers who had a 'professional' ethos with a new expanded base of programmers who, in the main, did not. And instead of managers being able to write their own programs in Cobol, as originally conceived by several of the language's designers, the new and growing workforce reconstituted the skills in software production and in many cases, from the point of view of capital, remystified the process of software production. This recomposition of skills, combined with the growing complexity of the software product itself, has robbed capital of the fruits of its successful initiative to such a point that today a leading consultant writes in a guide to computers for company directors:

> Software costs are, in strong contrast to those of hardware, rising because the preparation of tailor-made software is essentially labour intensive and, despite their keenness for other people to change their ideas, systems analysts and

> programmers are notoriously conservative in their own attitudes to fundamentally new and more productive methods.[12]

The composition of this 'conservative' labour force is in marked contrast to the profile of the microelectronic production and assembly worker we have given above. In the first place the only international division of labour that does exist is between metropolitan countries, the majority of whose software production is done in the US. Secondly, only 23% of programmers and 10% of systems analysts are women. Thirdly, a large percentage of programmers, about 35%, have between three and five years experience in their jobs. Within the metropolitan countries the concentration of software production is focused on the high-wage areas such as London and the Home Counties in the UK, whereas microelectronic hardware production has been drawn to the low-wage areas such as South Wales and Scotland. About 45% of programmers and analysts are in London and the Home Counties, and only 27% in the North, North East, East Anglia, Scotland and Wales.

In the cutback programmes of 1973 and 1974, management hacked away at training in attempts to diminish 'overheads'. As a result the number of trained and experienced programmers now available is not adequate for the task of developing software in the volume that capital considers necessary. Programmers are therefore, according to a recent survey, in a position to require between £500 and £1,000 extra a year before they will be persuaded to change jobs, all other things being equal.[13] But continuing attempts to cut back on 'overheads', variable capital, have meant that the majority of computer centre managers are faced with staff shortages because they will not pay the price determined by the labour market. But the initiative that capital is trying to mount — with a further division of labour, an enforced separation of the worker and the product, as well as the introduction of industrialised methods — has in general forced average wages downwards for software producers.[14] A report on salaries published in 1979 said:

> Computer staff appear to be losing the premium built up in the late 1960s and early 1970s . . . the more average performer, whose Cobol and Fortran alchemy is less essential as new machines, with much easier languages, become available, is

finding his salary drifting down to average levels.[15]

Programmers in the UK therefore start at about £3,000 a year, rise through the average of £4,250 and may, if they are 'good candidates who update their skills' as the report put it, reach the £6,000 a year level.

Office Automation

As with all trends towards the automation of existing work processes — including the very production of computer hardware and software — the initial areas of impact in the office are those that have already been, or can easily be, structured into repetitive, predictable and quantifiable tasks. Once this decomposition has been achieved by capitalist management, the functions can be recomposed by the injection of machinery and management intervention at key points through which labour will be controlled in the production process.

Management intervention is planned and co-ordinated in the office both in an individual sense of one firm and in a social sense of the state. Now the office has become a major bottleneck in the process of capital accumulation and has, therefore, become a target for automation in this decade. The bottleneck is two-fold: the productivity of office workers has not risen to the same extent as that of industrial workers, and capitalist management has detected an insubordination among office workers that challenges capital's control.

Within Western capitalist economies the number of office staff has grown 45% in the last ten years, while the total population of workers has risen only 6%. As the productivity of industrial workers has increased by at least 80%, that of clerical workers has only gone up 4%.[16] Within production, work has been formalised and automation has proceeded to the points at which it is bound by worker resistance and the survival of low labour cost areas. Within the office, work has been formalised in such activities as accounting, invoicing, calculation of salaries and recording of stock levels, but there is little automation beyond the traditional areas of electronic data processing that developed in the mid 1960s.

The most highly structured work, a typist's, is already under attack by the word processor. Instead of going over the ground investigated by Barker and Downing[17] we shall look at those office activities that are less structured, where work consists of exchanging ideas and reaching decisions.

The unstructured activity of the office takes up something between 50% and 70% of the time spent on office activity. The labour cost in Britain to perform these unstructured activities will be around £82.6 billion in 1980, according to projections from the UK Central Office of Information. While the cost of the workforce of clerks, who perform structured work, has risen 100% since 1975, the cost of the rest of the office workforce has risen nearly three-fold. The increase in cost for office workers performing structured tasks has been kept down by the extensive application of computer-based systems such as stock control and payroll which have expanded rapidly since 1969.

In the US the wages of clerical grade workers amount to only 27% of the total cost of running the office. The remaining $585 billion a year represents the costs of managers, professionals and other 'knowledge' workers.

". . . I find myself sitting here waiting for it to cross its legs . . ."

Two significant contributions have been made recently to the development of a capitalist strategy aimed at overcoming this relatively weak position: a senior executive in the Italian office products company Olivetti has gone a long way towards stating the problem in such a way that solutions will be systematically adopted, and IBM has finished research work on two prototype computer-based office systems.

Franco de Benedetti, joint managing director of Olivetti, made it quite clear where he located office automation:

> The problems which are usually associated with office automation make up the final chapter of a story which began with the industrial revolution, that is, with the supreme assertion of capitalist production over all preceding types of production and over all those that still exist today . . . Data processing is a continuation of a story which began with the industrial revolution, which incorporates the development of abstract methodologies within the development of technologies. Information technology is basically a technology of co-ordination and control of labour force, the white collar workers, which Taylorian organisation does not cover.

It is not only the lower rungs of management that 'exchange ideas and reach decisions' but also those many categories of 'professional' workers which have grown along with the development of the office — design engineers, planners, workers with professionalised skills attached to corporations. De Benedetti says of these offices: 'There is little routine in these offices where the work consists of meetings, exchange of documents and information. Any result achieved through a valid contribution from people involved in continuous interaction, implies unstructured activities.'

The information exchanged may be itself structured in phone calls, messages and letters, but the interaction between these types of information has not been procedurised. The role of the female secretary has been essential to increasing the productivity of these unstructured activities. The typewriter, the phone, filing cabinets and photocopiers are the tools that have allowed her to take the information from her 'principal' and make it intelligible to others, to transmit it to them and to receive their transmissions. The 'principal' is dependent upon the secretary for his or her effectiveness. 'This interdependence is often a negative factor which creates obstacles in the individual's work and reduces the efficiency with which the office as a whole pursues its objective.'

	£ Billion	
STRUCTURED WORK	1975	1980
Clerks	6.9	13.8
UNSTRUCTURED WORK		
Secretaries	3.2	6.4
Professional/Technical	12.6	32.2
Managerial/Administrative	17.0	44.0
TOTAL	39.7	96.4

Cost (£ Billion) of Office Workforce 1975/80
Source: U.K. Central Office of Information

In sketching out the way to break this interdependence, de Benedetti paints a picture in which the secretary has disappeared:

> Each office member should be provided with specific office tools to reduce interdependence and increase interaction; the employee can directly type the reports he is asked to compile, the manager returning to his office finds messages from people

> who want to get in touch with him (thanks to electronic mail), each office member is able to find electronically-filed material by himself; it is possible to organise meetings between people in different localities through teleprocessing techniques.

Just how some of this may be achieved has been outlined by IBM who found, when designing a prototype office communication system, that secretaries spent their working life doing the following:[18]

ACTIVITIES	AVERAGE PERCENT OF TIME
Writing	3.5
Mail handling	8.1
Bulk envelope stuffing	1.4
Collating/sorting	2.6
Proofreading	3.9
Reading	1.7
Typing	37.0
Telephone	10.5
Copying or duplication	6.2
Conferring with principals	4.3
Taking shorthand	5.5
Filing	4.6
Pulling files	2.8
Keeping calendars	2.6
Pick-up or delivery	2.2
Using equipment	1.3
Other	2.0

There is little in this list that could not be done automatically by the office system outlined by de Benedetti. As a move towards this goal the IBM prototype project discovered that upper level management appeared to shun administrative work and concentrate on communications work, but they could only do this with the co-operation of subordinates, mostly secretaries, who did the work for a secretary's, and not a manager's, salary.

> Even so, there was evidence that even more work could have been delegated if the proper people or systems were available. When asked if there were tasks that they did now that others could do for them, 51% of the principals indicated that they had one or more such delegatable tasks. At the time of this study principals were spending 14% of their work month doing tasks that, in their opinion, others could do for them. Our analysis of these tasks showed that many of them could be done by less highly compensated people and involved such activities as copying,

> calculating, assembling data, typing, covering telephones and filing and retrieving documents.

IBM's research depended upon a model of the office as a hierarchy of command built up of different layers, often visually represented as a pyramid with the board at the top and copy typists at the bottom. Even at the upper levels, which include professionalised workers, the more routine activities have been identified as already structured and hence ripe for delegation to machines.

But at crucial points within these hierarchical layers, radical breaks in composition occur — below which only the most menial decisions can be made, and below which the sex, race and class of the workers on one level differ from those above. Where these radical breaks occur, for instance between secretary and copy typist or between board members and line managers, status of employment changes significantly. This is often a reflection of the extent to which women or minority men have risen up the pyramid.

As office automation has been posited by the IBM research, successive layers of this hierarchy, from the base up, will be subjected to management 'science' analysis, the structured activities automated and the unstructured ones decomposed, in time, into a set of discrete and structured tasks. The IBM research does not recognise the differing composition of the different layers but instead presents a layer as composed of the same material as each other layer — just living labour. De Benedetti, however, does recognise the varied composition of the work force in the different layers.

> Given the high percentage of female personnel employed in the unstructured sector of office work — as secretaries — it should be remembered that this psychological reality is being profoundly changed by feminism.

There is also the possibility, he says, 'although no one in the business community wants to talk about it', that extensive white collar unionisation will occur as the 'private' relationships between the executive and his secretary come under public scrutiny.

Keeping the pyramid model in mind, we can see where the greatest strains in the hierarchy of the office will arise as a result of office automation achieved with microelectronic devices. First, strain will be felt where there is a radical difference between the composition of one layer and another. The command the layer above has over the layer below will begin to weaken even further with the undermining of the notion that the future for each individual in that layer lies in going up into the next layer. Where there is a radical difference between the composition of the two layers this notion is already almost non-existent.

Second, the strain in a layer will be experienced when the composition of that layer changes as part of the process of 'reorganisation' before or in parallel with the introduction of automation. Third, machinery will begin to drive the pace of the office to a much greater extent than it does at the moment, and to the point where there will be little distinction between the pace of a worker on a computerised numerical machine tool and a worker in the office at a microelectronic-based clerical work-station.

Mining Automation

In the coal pits of Great Britain a skilled group of workers who have considerable control over production is faced with extinction by the application of minicomputers and microelectronic devices. The National Coal Board (NCB) plans to spend £10 million in 1980 introducing the Mine Operating System (Minos) in 24 pits.[19] Face workers who currently control shearing and loading machines underground make up 20% of the pit workforce but their relative political strength among miners is shown by the fact that they hold about 80% of all union offices in pit branches. At the typical coal face today a team of half a dozen face workers control the cutting machines, Anderton Shearer Loaders, and power supports. Working on three rotating shifts, these face workers are supported by two to four electricians and fitters. Two deputies are the only direct expression of management control below the surface and, anyway, they are supposed to be primarily concerned with safety. Speeding up work could not be accomplished by management through direct supervision; hence the introduction, against some fierce opposition, of the piece work rates in 1978. But this financial level of subordination is only a half way stage. The Minos system will subordinate labour to machinery and shift control of production above the surface to management. The cutting machines and power supports will be fitted with microelectronic processors and memories linked back through minicomputers to a control room on the surface. Control room operators will be unskilled, and will need minimum training as the controller's interface with Minos will be a simple push-button command system using a colloquial interactive language. Behind this primary level of control there is a secondary management computer, complete with terminals in managers' offices which will provide access, restricted to managers, to production information and performance data.

The NCB is therefore planning to place automated machinery between the worker and the extraction process, breaking that skilled relationship between miner and materials and reconstructing it only when machinery has been placed between the two.

An important part of the NCB's plans for automation is a system of computerised controls for coal preparation plants on the surface. These plants wash and grade coal before it is transported to the user. A few of the smaller coal preparation plants have already been semi-automated and work has just finished on one of the larger plants at the South Kirby colliery in Yorkshire, at a cost of £18 million. Through the increase in productivity that automation brings, this new plant will receive coal from surrounding pits that used to be prepared and shipped at the pit head. To this job loss at the other preparation plants must be added the effects of de-skilling among the remaining coal preparation workers at South Kirby.

The new plant is a closed-loop system where computers, not the human operators, control the work process. Workers merely monitor its operation and, in their turn, are monitored by another part of the system run on a different computer which feeds information to management. The NCB has regraded the jobs on the plant downwards because, it says, less skill is needed to control the system. The National Union of Miners' branch at the pit has responded with demands for upgrading and more money.

Both below and above ground the miners and surface workers will be subordinated to the machinery instead of having control over the processes. The computer systems in Minos physically extend the control which management has, because the devices on the shearing and loading machinery will be constantly feeding data back to the surface and, eventually, receiving commands in the form of electronic signals from the control room, also on the surface. The total number of workers needed to run these plants will be less. The Minos application is therefore aimed by the NCB at two targets at the same time — cutting the work force and increasing the command which management has over the remainder of the workers.

It is worth adding that automation at the coal face is one jaw of a vice which is being designed to squeeze this militant sector of the working class, the other jaw being the development of nuclear power as an alternative energy source. In crisis periods the availability of nuclear power could remove the key political position of miners: they could no longer 'hold the country to ransom'.

New Jobs and Skill Shortages

As the effects of microelectronics begin to be discussed in public, workers are promised that new industries based on microelectronics will emerge to take in the labour 'shaken out' from older industries. These new microelectronic-based industries will be mostly export-orientated and will grow at above-average rates. One such industry which fulfills all the criteria held by these 'technological optimists' is the UK radio communications, radar and navigational aids industry whose sales have grown in the past three years by 16% per annum.[20] Since 1975 this industry has taken on 5,000 new workers to boost its total employment to around 94,000. This growth, based on a rapidly expanding export market, has not been matched by training and hence the industry has outgrown the ability of the skilled job market to fulfill its need for trained workers. A report into the industry by the National Economic Development Organisation said in 1979,

> If grossed up these shortages imply a total shortfall of about 9,100 people, of which some 2,700 are craftsmen and operators and 6,400 are qualified and technical staff. Moreover changes in product characteristics and manufacturing and design methods are already, and will continue, to have manpower effects. The level of skill is increasing and the type of skill needed is changing as more advanced technologies are adopted.

An electronic engineering survey conducted in late 1977 identified the greatest shortages in the areas of computer programming and systems analysts (34% shortfall), technical authors (26%), draughtsmen (9%) and craftsmen and machine operators (8%). The NEDO report on the industry, which was drawn up by a working party including representatives from the

APEX, ASTMS and EETPU unions[21], did not propose a crash training course to bring new skills to the workforce. Instead it said: 'Some of these shortages could be alleviated by increased automation and the deskilling of routine operations in the field of design, manufacture and test by the use of computer aided design, numerical control machinery and automatic testgear.' It is therefore proposing to substitute for human skill machinery based on modern microelectronics which will rob a large section of workers, 9,100 in this industry alone, of the chances of becoming skilled workers with their skills recognised by capital. The substitution of microelectronic-based devices for labour within production, therefore, will not only occur in the older industries where formal working class organisation is strong, but also in younger industries which provide the only hope of job creation within a generally contracting capitalist economy. In alliance with a trade union leadership fixated solely by wage struggles, capital is trying to transfer the initiative it has gained over workers in the microelectronic hardware production industry into a general initiative against the working class in other industries. In the production of new commodities based on microelectronics, capital aims to press its initiative to the point where the working class in those production processes has no chance to reconstitute the skills it has in production. In short, microelectronics provides part of the strategy of capital to re-establish its hegemony over production.

Conclusions

It is often argued that the types of applications we have analysed in this chapter are only an abuse of technology. Technology's real potential is to free us all from tedious and unhealthy work environments, increase our productivity and therefore our wages, and expand the amount of leisure time at our disposal. Any application — industrial, clerical or domestic — that does not 'release' this potential is misguided and abusive. After all, the argument goes, was not the work of the original designers of the first computer, and of programmers deeply immersed in complex systems that stretched their skills, both interesting and rewarding? And what of the beneficial applications of microelectronics — computer controlled X-ray scanners aiding the speedy diagnosis of cancer while educational material can be beamed round the world by digitally controlled television satellites? Such beneficial applications certainly do exist. But some indication of the place they hold in capitalist priorites is given when looking at capitalist research and development funding.

In the United States between 1974 and 1975 the state spent 50% of its total research and development funding of over $17 billion on 'national defence' and only just over 10% on health. Years of polemic and campaigning by reformers have failed to shift substantially this ratio. We can only conclude, therefore, that the priorities are systematic, and recognise that the same is true for the application of one of the greatest fruits of research and development — microelectronics. Further than that, we began deliberately with an analysis of chip and software production — exposing the myth that the imaginative use of skills is the prerogative of the production industry itself. Far from it: any exercise of skill in the production process is now under attack as managers advance the slogan that design and programming must change from an 'art form' into a science.

The new technology is not, therefore, a movable feast which may be 'used' or 'abused' at will. It is as much a product of capitalist society as war, famine, racialism, sexism and economic exploitation. Equally, technology not only bears the mark of the society in which it was developed, it is also used as an active agent in that society by those who have the means to continue technological development to the point at which it becomes a useful tool in the battle to retain and increase control over society. Women on the assembly lines in South East Asia do not determine the applications for microelectronics in the metropolitan countries where computers and automation are most advanced. Equally, programmers are being given less and less information about the actual application they are working on as the methods of structured programming allow management to segment one part of the program or system from another so that no single programmer can get an overall view.

In the research and development plants of the manufacturers, management presents mental workers with technical choices. Only when it comes to the choice, by management, on a specific application do the underlying political, economic and social aspects of the technology come to the surface in the discussion. Then the criteria are profitability and productivity. Management wants to keep for itself the right to decide where, when and how the developments of microelectronics shall be applied. They therefore show to the outside world only the technical aspects of the technology, presenting it as a neutral tool.

We would, however, not have this technology without the exploitation of production workers in California and South East Asia. And even if it were decided that the benefits to the rest of us outweigh the exploitation of these workers, this technology could not be applied in just any way that was seen fit, because the form of the technology is determined by the society which gave it birth. The object of production within capitalism is the creation of surplus value which will go as profit to the capitalist. The aim of production is not the manufacture of socially useful products nor the development of work practices that will enrich and express the humanity of workers. The application of microelectronics, as well as their development and manufacture, is indissolubly bound up with the capitalists' aim of making profits. It is true that we, as workers, can struggle for applications that make our places of work healthier and safer and for applications that extend our skills instead of stripping us of our means of livelihood. But we must always bear in mind that any such applications will have to be won against fierce resistance and that, further, such applications will be used against us if capital can find any possible way of doing so. From a safety point of view it would, for instance, be useful to know exactly where all the miners are in a pit and what they are doing. But the National Coal Board's Minos system has not been built to enhance the safety of the work force, and even such a limited monitoring system showing where every miner is would rapidly become a tool to increase management's control. Hence, workers often breach safety regulations precisely because they act as an impediment to achieving a decent wage when working on piece rate or because they allow management too much supervision.

There are many areas that we have not considered — not because car production, telecommunications maintenance, clothing manufacture, oil processing and the rest are not important, but because the five applications we have looked at throw light on crucial aspects common to all.

First, in the production of the technology itself, international and sexual divisions of labour are used to extract the greatest profit from, and exert the greatest control over, workers. Second, in the production of the programs to run on the devices, mental workers are under the same pressure, facing the

A NO OPTION FUTURE ?

NO THE 35-HOUR WEEK GETS THUMBS DOWN FROM MINISTRY

By PETER McHUGH

HOPES of a 35-hour working week were blasted by a Government report yesterday.

IT REJECTED union claims that a cut in the present 40-hour average week would create more jobs.

INSTEAD it came out in favour of slashing overtime as a means of reducing unemployment.

Now, Employment Secretary Albert Booth faces angry union reaction to the report.

Albert Booth . . . facing a row with unions

agree that, on paper, a 35-hour week could take 500,000 off the dole queues.

"vicious circle" of higher production costs, less competitive prices and, finally, damage employment prospects again.

The same applies to extending annual holidays, said the report.

Resent

The most significant effect on jobs could come from cutting overtime, it claimed.

At least a quarter of Britain's manual workers are working more than eight hours extra each week.

But the researchers

THERE IS NO CERTAINTY ABOUT JOB LOSS IF WE DO APPLY MICRO-ELECTRONIC TECHNOLOGY. THERE IS ABSOLUTE CERTAINTY ABOUT JOB LOSS IF WE DO NOT.

Employment Minister, Albert Booth.
Sunday Times 17 - 9 - 78

same battles as manual workers. Third, technology does not provide a clean and neat solution for the problems capital has with its workforce. As workers are deskilled and laid off in one industry, so the area of software production becomes a crucial point of struggle between capital and labour over productivity and control. As the section on software production shows, capital cannot successfully eject labour as a whole from the production process. In short, the struggle between capital and labour is reproduced in another area, just as capital thought it might be over once and for all.

Fourth, the office in general, not only the places where software producers work, will become a battle ground over productivity and control. The more menial functions are now under attack and the preliminary work is going on to soften up other areas of office work for their eventual replacement by microelectronic-based systems. Fifth, where capital is strategically weak — where its control of the labour force is low and where workers in that labour force have provided a leadership for other workers — these areas, such as mining, are also under attack with large capital investment programmes.

Sixth, many trade union leaders call for higher levels of capital investment in British industry which, with the dominant role that microelectronic technology now plays, can only mean more automated processes.

In doing this they are effectively calling for an extension of the process of deskilling and increased management control that is part and parcel of capital's new technology. Even in those areas that are presently growing in output, demands for a new skilled workforce are met with proposals to automate more and more of the processes of design, coordination and manufacture.

Notes

1 For example Ray Rothwell and Walter Zegveld in *Technical Challenge and Employment*, Frances Pinter, 1979, deal with past and future trends of employment and unemployment in mature industrialised societies with case studies on the impact of technological change. Christopher Hill and James Utterback have edited a selection of papers called *Technological Innovation for a Dynamic Economy*, Pergamon Policy Studies, 1980, in which Clinton Bourdon reviews classical approaches in 'Labour, Productivity, and Technological Innovation: From Automation Scare to Productivity Decline', pp.222-54.

2 *Annual Report*, Intel Corporation, 1979, declares a turnover of $662 million and a profit before taxes of $149 million on the work of some 14,300 workers backed by assets totalling $500 million.

3 *The Electronics Industry in Malaysia*, US Department of Commerce, August 1978. A report to the Federal Government from the Commercial Section of the United States Embassy in Kuala Lumpur, available from the US National Technical Information Service.

4 Rachael Grossman's 'Silicon's Ugly Secret', *Computing*, September 6, 1979, reports on the personnel management techniques used by microelectronic corporations to manipulate women in South East Asian production plants. Grossman and others report on the *Changing Role of S.E. Asian Women* in a special joint issue of *South East Asia Chronicle* 66 (Jan-Feb 1979)/*Pacific Research* IX, 5-6 (July-Oct 1978).

5 A. Sivanandan, 'Imperialism and Disorganic Development in the Silicon Age', *Race and Class*, Volume XXI, Number 2 (Autumn 1979).

6 Quantum Science Corporation estimate presented at the Annual Strategy Conference, June 1979. Not published.

7 *Annual Report*, Intel, 1978.

8 Quantum Science estimate.

9 Gene Bylinsky, 'Here Comes the Second Computer Revolution', *Fortune*, November 1975.

10 An excellent example of the type of tactics being urged on management is a recent handbook by Edward Yourdon, *Managing the Structured Techniques*, New York, Prentice-Hall, 1979.

11 J R M Simmons, *Leo and the Managers*, MacDonald, 1962.

12 *The Director's Guide to Computing*, prepared by the Institute of Directors in collaboration with the National Computing Centre, revised edition October 1979.

13 'DP Staff Loyalty Shock', *Datalink*, 22 October 1979.

14 For a general analysis of the effects of structured programming see: Joan Greenbaum, *In the Name of Efficiency*, Philadelphia, Temple University Press, 1979; Philip Kraft, *Programmers and Managers*, New York, Springer Verlag, 1977.

15 'Wages Are Drifting Down', *Computing*, November 8, 1979, page 5.

16 Figures presented by Franco de Benedetti, Olivetti, to the *Financial Times* conference but not published.

17 Jane Barker and Hazel Downing 'Word Processing the Transformation of the Patriarchical Relations of Control in the Office', *Capital and Class* 10 (Spring 1980), pp. 64-99

18 *IBM Systems Journal*, volume 18, number 3 (Autumn 1979), pp. 402-456.

19 Dave Feikart, 'Of Man and Minos', *Computing*, November 22, 1979, p. 31.

20 National Economic Development Council Radio Communications, Radar and Navagational Aids sector working party interim report, April 1979.

21 Association of Professional, Executive, Clerical and Computer Staff (APEX); Association of Scientific, Technical and Managerial Staffs (ASTMS); Electrical, Electronic, Telecommunications and Plumbing Union (EETPU).

CSE Books

CSE Books was founded by members of the Conference of Socialist Economists to promote the practical criticism of capitalism which the CSE as a whole is committed to and to facilitate wider participation in the debate and analysis going on in the CSE. Rather than forming ourselves into an academic editorial committee which sits in judgement of authors and in ignorance of readers, we want to engage politically in current debates and struggles. By coordinating with CSE activities in general, by publishing *Head & Hand: A Socialist Review of Books*, and by organising dayschools on issues thrown up by our own publications, we hope to narrow the gap which exists in bourgeois society between the producers and consumers of books.

For further information on CSE Books titles and the CSE Bookclub, write to 55 Mount Pleasant, London WC1X 0AE.

The Conference of Socialist Economists

The Conference of Socialist Economists was formed in 1970. Since that time there have been many changes. CSE is committed to development of a materialist critique of capitalism in the Marxist tradition. The membership of CSE now covers a broad spectrum of political and research activities which generates a wide-ranging debate, for CSE is as far as possible, unconstrained by the traditional academic divisions of intellectual labour into, say, 'economists', 'politics', 'sociology' or 'history'.

Instead the groupings are around the CSE working groups. Currently, groups actively working on material are the Ideology Group, Housing, Capital and the State, State Economic Policy, Capitalist Labour Process, Political Economy of Women, European Integration, Health and Social Policy. There is a Labour Process Historians' Group and there may at different times be other groups in operation. Groups are in various parts of the country.

For further information on membership, write to CSE, 55 Mount Pleasant, London WC1X 0AE.

HEAD & HAND: A SOCIALIST REVIEW OF BOOKS

Head & Hand is the only English-language periodical devoted to reviewing a wide range of publications from a socialist perspective. It provides an alternative to its liberal counterparts and to the scattering of socialist reviews in more specialist journals. It carries highly readable, well-informed and often provocative reviews: novels, politics, history, trade unions, children's literature, feminism, health, community publishing, socialist theory & practice, international affairs, nuclear power, Ireland etc. As well as essay reviews, it also contains short reviews and notices, reading lists and news from CSE Books and other publishers. 24 A3 pages.
Subscription £2.50 (overseas £3) for 3 issues per year.

Other New Titles from CSE Books

Microelectronics: Capitalist Technology and the Working Class, CSE Microelectronics Group.
Now that the post-war boom has ended, microelectronic technology will be the key element in restructuring the factory and the office as capital attempts to lay the basis for a new period of accumulation. Cheap mass-produced processors etched onto tiny chips will enable management to re-organise jobs in ways that undermine or exclude workers' collective power. This book examines capital's strategy for reasserting its control over the labour process in different sectors of the economy as well as the complementary role of the state in promoting the development and application of microelectronics. 152 pages illust.
Hb 0 906336 16 3 £8/Pb 0 906336 17 1 £2.95.

Living Thinkwork: Where Do Labour Processes Come From? Mike Hales.
Mike Hales describes his experience doing operations research at ICI, where even mental workers learn that ultimately 'You're not paid to think'. Through an account of 'scientific' work in a capitalist firm, his book shows the place of knowledge-production in the politics of management. A concrete intervention in Marxist theory of the labour process, this book is also a document in the history of the 'class of '68', exploring the contradictory social relations between theory and personal experience, theory and practice, and academic and industrial work. 192 pages illust.
Hb 0 906336 14 7 £10/Pb 0 906336 15 5 £3.50.

Northern Ireland: Between Civil Rights and Civil War, Liam O'Dowd, Bill Rolston & Mike Tomlinson.
This book is the first major study to document the origins and nature of Direct Rule in Northern Ireland, particularly the social-democratic model which Britain has attempted to superimpose upon the Orange State. The authors argue that Direct Rule has not been overcoming the notorious sectarianism of Stormont but has instead reconstituted class sectarian relations more subtlely within the new state institutions. Their argument is illustrated through detailed studies of the economy, trade unions, local government, housing, community politics and repression. 224 pages illust.
Hb 0 906336 18 X £12/Pb 0 906336 19 8 £3.95.

The Alternative Economic Strategy: A Labour Movement Response to the Economic Crisis, CSE London Working Group.
This book argues that there is an alternative to Tory monetarism: a socialist transformation of the economy.
Hb 0 906336 22 8 £7/Pb 0 906336 23 6 £2.50

Also Available from CSE Books

Technology and Toil in Nineteenth Century Britain, ed. Maxine Berg. A collection of documents designed to illustrate the neglected early history of deskilling, technological unemployment and assembly line alienation. The fifty-one selections reveal differing practices in a wide range of industries and provide a context in which the better known analyses of work and technology left by Babbage, Ure, Owen and Marx may be analysed. 250 pages illust.
Hb 0 906336 02 3 £10.00 / Pb 0 906336 03 1 £3.50.

Economy and Class Structure of German Fascism, Alfred Sohn-Rethel.
A classic Marxist study of the founding of the Nazi state in the early 1930s. Sohn-Rethel managed to infiltrate the office of the Mittel-europaeischer Wirtschaftstag, a business institute whose members included representatives of every significant section of German finance capital, and the book is based on his experiences as an editorial assistant on the MWT's newsletter. 160 pages.
Hb 0 906336 00 7 £5.95 / Pb 0 906336 01 5 £2.50.

Struggle Over the State, CSE State Group
Building on the current upsurge of Marxist interest in the state, this collectively produced work combines recent theoretical progress with informed and up-to-date description of the restructuring of the UK state in the crisis of the 1970s. 144 pages illust.
Hb 0 906336 13 9 £5.95 / Pb 0 906336 12 0 £2.50

Value: The Representation of Labour in Capitalism, ed. Diane Elson. Marx's theory of value has been a controversial and constant matter of debate since the time it was first published. This collection of essays focuses on some of the more difficult and neglected concepts at the very beginning of *Capital*: abstract labour as the substance of value; the relative and equivalent forms of value; exchange value as a necessary mode of expression of value; the commodity as a symbol. The book also contains a useful annotated bibliography of all Marx's references to value. 192 pages.
Hb 0 906336 07 4 £12.00 / Pb 0 906336 08 2 £4.95.

A Workers' Enquiry into the Motor Industry, IWC Motors Group. Examines redundancies, speed-ups, payments systems, automation, health and safety and the international reorganisation of capital investment—all from the point of view of labour. A4 size, 102 pages illust.
0 906336 05 8 £2.25.